Global Perspectives on Health Geography

Series Editor

Valorie Crooks, Department of Geography, Simon Fraser University,
Burnaby, BC, Canada

Global Perspectives on Health Geography showcases cutting-edge health geography research that addresses pressing, contemporary aspects of the health-place interface. The bi-directional influence between health and place has been acknowledged for centuries, and understanding traditional and contemporary aspects of this connection is at the core of the discipline of health geography. Health geographers, for example, have: shown the complex ways in which places influence and directly impact our health; documented how and why we seek specific spaces to improve our wellbeing; and revealed how policies and practices across multiple scales affect health care delivery and receipt.

The series publishes a comprehensive portfolio of monographs and edited volumes that document the latest research in this important discipline. Proposals are accepted across a broad and ever-developing swath of topics as diverse as the discipline of health geography itself, including transnational health mobilities, experiential accounts of health and wellbeing, global-local health policies and practices, mHealth, environmental health (in)equity, theoretical approaches, and emerging spatial technologies as they relate to health and health services. Volumes in this series draw forth new methods, ways of thinking, and approaches to examining spatial and place-based aspects of health and health care across scales. They also weave together connections between health geography and other health and social science disciplines, and in doing so highlight the importance of spatial thinking.

Dr. Valorie Crooks (Simon Fraser University, crooks@sfu.ca) is the Series Editor of Global Perspectives on Health Geography. An author/editor questionnaire and book proposal form can be obtained from Publishing Editor Zachary Romano (zachary.romano@springer.com).

More information about this series at https://link.springer.com/bookseries/15801

Caroline Barakat • Fatme Al Anouti
Editors

Adolescent Mental Health in The Middle East and North Africa

Springer

Editors
Caroline Barakat
Faculty of Health Sciences
Ontario Tech University
Oshawa, ON, Canada

Fatme Al Anouti
Department of Natural Sciences and Public Health
Zayed University
Abu Dhabi, United Arab Emirates

ISSN 2522-8005 ISSN 2522-8013 (electronic)
Global Perspectives on Health Geography
ISBN 978-3-030-91789-0 ISBN 978-3-030-91790-6 (eBook)
https://doi.org/10.1007/978-3-030-91790-6

This Springer imprint is published by the registered company Springer Nature Switzerland AG
The registered company address is: Gewerbestrasse 11, 6330 Cham, Switzerland

Acknowledgments

We would like to thank Susan Yousufzai and Moatassem Kenaan for their support and assistance in coordinating this book project. Many thanks to the contributing authors who continue to demonstrate deep enthusiasm and commitment to their research journey and academia. We are endlessly indebted to our (adolescent) children and thank them for their understanding and support, especially as we shifted a lot of our time and attention to completing this book. We dedicate this book to all adolescents, especially those from the Middle East and North Africa. You and your perseverance are the sources of inspiration that keep us going.

Contents

Contributors

Siwaar Abouhala Advancing Arab American Health Network and Allies (AAAHNA) Research Group, Tufts University, Boston, MA, USA

Akram Abusalah Nursing Sciences, Palestine College of Nursing, Gaza Strip, Palestine

Rasmieh Al-amer School of Nursing, Isra University, Amman, Jordan

School of Nursing and Midwifery, Western Sydney University, Penrith South, NSW, Australia

Fadwa Alhalaiqa Faculty of Nursing, Philadelphia University, Amman, Jordan

Mohammed AlKhaldi Faculty of Medicine, School of Physical and Occupational Therapy, McGill University, Montreal, QC, Canada

Council on Health Research for Development, COHRED, Genève, Switzerland

University of Basel, Swiss Tropical and Public Health Institute, Department of Public Health, the unit of Health Systems and Policies, Basel, Switzerland

An-Najah National University, Faculty of Medicine and Health Sciences, Nablus, Palestine

Hala Allabadi Public Health and Epidemiology, An Najah National University, Nablus, Palestine

Fatme Al Anouti Department of Natural Sciences and Public Health, Zayed University, Abu Dhabi, United Arab Emirates

Caroline Barakat Faculty of Health Sciences, Ontario Tech University, Oshawa, ON, Canada

Jennifer Dabis Juzoor for Health and Social Development, Al-Bireh, Palestine

Alaa Galadari Sheikh Khalifa Medical City, Abu Dhabi, United Arab Emirates

Aya Ibrahim United Nations Economic and Social Commission for Western Asia, Beirut, Lebanon

Rima A. Kashash Palestine Red Crescent Society/Lebanon Branch, Akka Center, Ghobeiry, Lebanon

Patricia Niland Accident Compensation Corporation, Auckland, New Zealand

Mohd Fadzil Abdul Rashid Universiti Teknologi MARA, Shah Alam, Malaysia

Asharaf Abdul Salam Center for Population Studies, King Saud University, Riyadh, Saudi Arabia

Justin Thomas Department of Natural Sciences and Public Health, Zayed University, Abu Dhabi, United Arab Emirates

Part I
Introduction

Adolescent Mental Health in the Middle East and North Africa (MENA): Where Are We and Where Do We Go from Here?

Caroline Barakat and Fatme Al Anouti

1 Introduction

Childhood and adolescence are time periods for human brain development and growth. Neuroimaging studies show that white matter volume and integrity increases throughout childhood and adolescence, with the frontal and temporal regions of the brain undergoing pronounced changes in adolescence (Tamnes et al. 2010). Decreases in frontal and temporal gray matter volume during adolescence result from environmental influences, as well as increasing white matter, the latter being essential as a transport mechanism of vital brain connections (Giedd et al. 1999). In parallel to these physiologic changes, marked cognitive changes include increases in abstract thought and reasoning (Piaget et al. 1977), improvements in IQ and working memory (Ostby et al. 2011), better problem solving and planning (Squeglia et al. 2013), and maturation of perspective taking (Sebastian et al. 2012). These changes suggest that the human brain exhibits heightened neural plasticity during adolescence, defined as the brain ability to adapt to internal or external changes. Environmental exposures and influences during adolescence determine these changes. For example, contextual neighborhood features such as the absence of opportunities for physical activity or an abundance of vaping stores influence adolescents to exhibit health risk behaviors such as sedentary lifestyles and drug use, which in turn influence the development of health outcomes, such as obesity and mental disorders. Fuhrmann et al. (2015) highlighted adolescence as a sensitive

C. Barakat (✉)
Faculty of Health Sciences, Ontario Tech University, Oshawa, ON, Canada
e-mail: caroline.barakat@ontariotechu.ca

F. Al Anouti
Department of Natural Sciences and Public Health, Zayed University,
Abu Dhabi, United Arab Emirates

C. Barakat, F. Al Anouti (eds.), *Adolescent Mental Health in The Middle East and North Africa*, Global Perspectives on Health Geography,
https://doi.org/10.1007/978-3-030-91790-6_1

period of development where brain changes may be particularly susceptible to environmental influences, such as drugs, social stress, or cognitive training.

Brain development and growth relate to mental health over the life course. Mental health encompasses a person's psychological, emotional, and social well-being (CDC 2018). More recently, mental health is recognized to be on a spectrum, as opposed to two ends of the spectrum, positive mental health and mental illness (Provencher and Keyes 2011). The World Health Organization (WHO) defines mental health as "a state of well-being in which the individual realizes his or her abilities, can cope with the normal stresses of life, can work productively and fruitfully, and can contribute to his or her community" (Galderisi et al. 2015). This definition raises many questions: for example, how can one determine that he or she is realizing his or her abilities? What constitutes coping? What is "normal" in relation to diverse life stresses?

Mental illnesses constitute the absence of or deficits in mental health. They are disorders or conditions that negatively affect a person's psychological, emotional, and social well-being. They tend to limit one's abilities to manage daily activities, and cope with life stresses. Additionally, mental illness often hinders productivity and positive thinking (National Alliance on Mental Illness 2017). There are diverse mental disorders that differ in relation to outcomes and symptoms. Anxiety is one of the most common mental disorders with a staggering prevalence estimation of 284 million people globally (Ritchie and Roser 2018). Depression is another common mental disorder with an estimated 264 million people suffering globally (WHO 2019). The latter exhibits extreme outbursts of sadness, unproductivity, and negative intruding thoughts. Moreover, there are behavioral mental illnesses like autism, attention-deficit/hyperactivity disorder, and childhood anxiety. With over 200 classified mental illnesses, each one is unique in its characteristics. Many of them have similar attributes like inability to perform normal life activities, low productivity, and loss of energy. However, some have distinctive symptoms; for example, depression might lead to suicidal thoughts, while bulimia leads to excessive food binging followed by trying to get rid of consumed calories through forced vomiting (Ritchie and Roser 2018).

It is unrealistic to remain mentally healthy for the entirety of one's life (Ronald et al. 2010). Most individuals experience disordered thinking, behavior, or emotions at some point in their lives. Wellness or well-being focuses on the positive aspects of mental health and stems from our ability to understand, accept, and embrace our capabilities to indulge in life experiences and opportunities. For instance, to improve mental well-being, one has to fulfill six dimensions of the wellness model, which encompass the emotional, occupational, physical, social, intellectual, and spiritual aspects of life (National Wellness Institute). The emotional dimension focuses on persons' potential to be aware of their feelings and those around them, in order to confront these feelings and grow emotionally. Equally important is having a profession or vocation that allows an individual to work toward a goal or contribute to a community, thus offering a sense of contentment and satisfaction. Vocational activities such as volunteering help the individual manage stress and release it through a passionate outlet. Maintaining physical health is pivotal to improving an individual's mental wellness and health; healthy balanced diets, regular physical activity, and sleep hygiene are essential to maintain physical health. Social wellness can be enhanced through maintaining a good level of social activities, such as socializing and focusing on healthy relationships with friends and family members. Remaining intellectually active and spiritually connected offer the benefits of agility, attentiveness, and purpose. Creative outlets such as reading or painting enhance intellectual awareness, while spiritual growth and involvement may offer guidance and direction (Hoebeke 2019; Roddick 2016).

The abovementioned six dimensions of wellness support the idea of multifactorial influences in relation to mental health and illness. This is particularly true for adolescents, who are subject to new experiences that influence and shape all six dimensions of wellness. During adolescence, individuals tend to negotiate a myriad of emotions and learn to understand and regulate their feelings; they feel the need to focus on their future career options and processes in order to achieve their career goals. Furthermore, the increased independence that adolescents experience allows them to make choices in relation to heath behaviors, such as food choices or physical activity patterns. Friends are very important during adolescence; thus, peer influence is a major factor that contributes to all six dimensions of wellness, particularly to the social, intellectual, and spiritual dimensions. Importantly, all six dimensions of wellness are influenced by diverse individual and environmental circumstances. Here, one simply has to conceptualize the holistic Health Onion Model (Dahlgren and Whitehead 1991; Marks 2002). This model takes account of a person's health as having many interconnected layers (like an onion). Individual factors that influence health are age, gender, hereditary, and lifestyle choices (e.g., healthy diet, exercise). Community, living and working conditions, and the socioeconomic, political, cultural, and physical environment also influence the health of an individual. Thus, important predictors of health are family, friends, housing, employment, education, religion, government, and healthcare systems. For example, in relation to mental health, while free medical care and services contribute to positive health outcomes, mental illness stigma is a sociocultural predictor that impedes well-being and hinders access to healthcare. In fact, despite estimates of 792 million people suffering from mental disorders, mental illness remains stigmatized (Ritchie and Roser 2018). Individuals that suffer from mental illnesses may not seek help and would shy away from seeking treatment, leading to underestimates of the prevalence of mental illnesses. For adolescents, the WHO estimates that 10–20% of adolescents and children experience a mental disorder or illness.

MENA is an acronym that refers to "the Middle East and North Africa." The region encompasses Algeria, Bahrain, Egypt, Iran, Iraq, Israel, Jordon, Kuwait, Lebanon, Libya, Morocco, Palestine, Oman, Qatar, Saudi Arabia, Sudan, Syrian Arab Republic, Tunisia, United

Arab Emirates, and Yemen (Fig. 1). The term "MENA" is very conveniently used by academic, social, economic, and international entities based on not only geographic proximity but also the relative homogeneity of the different countries in terms of culture and religion. The World Bank and United Nations lately started using MENA term to refer to all countries that lie between Iran and Morocco. It is

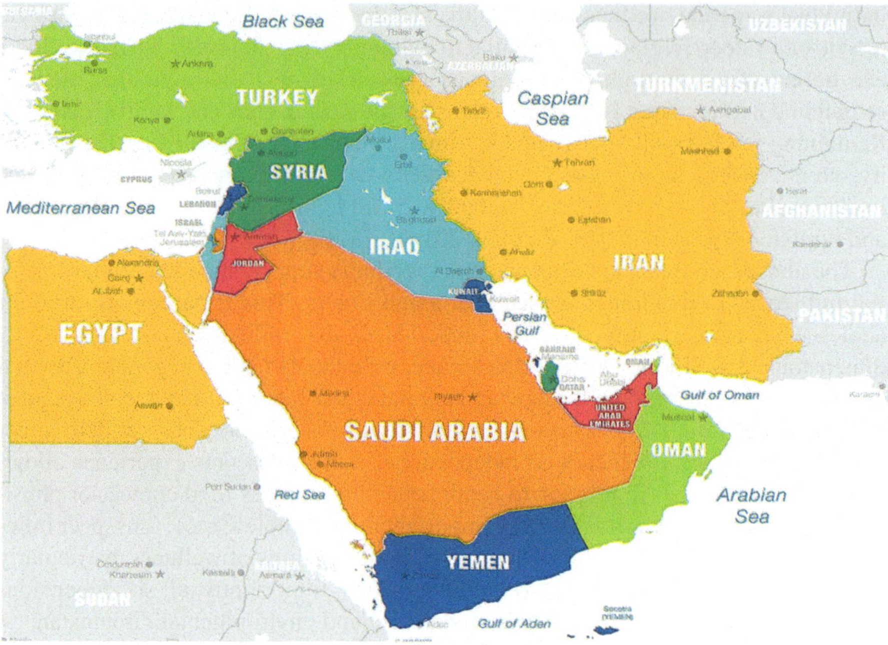

Fig. 1 Map of MENA. (Source: https://www.worldatlas.com)

worth mentioning that the acronym has differences in terms of included countries. For instance, there are times when MENA also includes Turkey (https://www.worldatlas.com/articles/what-are-the-mena-countries.html).

MENA region contributes to approximately 4.5% of the world's gross domestic product (GDP) and is conveniently accessible to international markets. Possessing 60% of the world's oil and 45% of the world's natural gas reserves, MENA has a global economic importance (WHO 2016). MENA has a relatively small population mostly living in middle-income countries and has one of the highest adolescent birth rates globally. Children, adolescents, and young adults account for almost half of the region's population. Since youth is an essential factor for change, the region has vast opportunities for economic growth (UNICEF 2019). Nonetheless, this subpopulation could be vulnerable and underprivileged because of several challenges that specifically prevail in MENA. For instance, MENA holds the highest youth unemployment rate in the world (Karamouzian and Madani 2020). Evidence is mounting regarding the high prevalence of mental health issues among adolescents from MENA given the numerous war conflicts, violence, and political instability in some countries like Syria, Iraq, Yemen, Lebanon, and Palestine (WHO 2019). The availability and access to health services and infrastructure along with political, cultural, and socioeconomic conditions (e.g., strict family expectations, unemployment, gender gap) are among the major determinants that shape the mental health of adolescents (AlBuhairan 2015).

The divergent conditions across countries in MENA lead to complexities and disparities in relation to adolescent mental health (AlBuhairan 2015). A remarkable contrast exists within MENA, as the region encompasses low-income countries afflicted by war, as well as countries with the highest per capita income like the Arab Gulf (WHO 2016). In politically unstable countries like Syria, Lebanon, and Palestine, adolescent refugees are marginalized and have limited access to health services and education, leaving such countries lagging in implementing effective prevention efforts and policies to overcome the burden of mental health (AlBuhairan 2015). On the other hand, while adolescents in the Gulf region have access to quality healthcare services and infrastructure, addiction and depression are among the most pressing mental health issues that the concerned governments are targeting for intervention (Al Makadma 2017). The Arab Youth Survey (2019) included questions related to "mental health" for the first time in its 11 years of publication. Results highlighted that 50% of youth respondents viewed mental illness negatively, thus echoing the importance of social and cultural attitudes toward mental health in the Arab world.

Research that investigated adolescent mental health in MENA is essential for the design of effective mental health interventions. Increased knowledge on the challenges and tensions that adolescents in MENA face is essential for mental health initiatives and policies and can guide in addressing individual, family, and sociocultural contexts, such as local values and beliefs and indigenous health practices. Research that translates adolescent perspectives into culturally appropriate interventions can provide guidance and resources for mental health practitioners (Al-Krenawi 2005).

This book consists of six chapters that cover various topics related to adolescent mental health in MENA. The first chapter outlines the importance of considering mental health as a major dimension of adolescent health status. An overview of evidence on the status of mental health of adolescents in MENA follows, leading to evidence of mental illness stigma that is rampant in MENA; thus, the third chapter is dedicated to exploring mental health stigma in Arab youths. The last three chapters of the book focus on policy and programs that target adolescent mental health in MENA, with a focus on youth-friendly public policies and future directions.

1.1 Current Status of Adolescent Mental Health Research in MENA

Health encompasses various dimensions, one of which is mental health, as emphasized in Chap. 1. Thomas and Galadari uncover diverse factors that influence the mental health of adolescents in MENA. They discuss the female body ideal of thinness and the fear of being "fat," and link this to eating disorders and mental illnesses. They also highlight various factors linked to depression, including financial stressors, stressful life events, female gender, lack of support, chronic illnesses such as diabetes and obesity, and history of eating disorders. Thomas and Galadari point out that the overall prevalence of depression ranges from 13% to 18% in MENA. Risk factors of depression include financial stressors, stressful life events, female gender, lack of support, chronic illnesses (including diabetes and obesity), and a history of eating disorders.

A review on the prevalence of depression among adolescents from MENA found rates as high as 86% in some regions (Al-Amer). Specifically, the author report that the prevalence of depression among adolescents ranges from 4.1% and 41.5% in Saudi Arabia, 17% to 32% in Egypt, 30% in Qatar, 17.5% in Dubai, 40% to 41% in Jordan, 13.5% to 43.55% in Iran, 86% in the Gaza Strip, and 63.4% in Iraq. Community studies across MENA region demonstrate that girls are two times more likely than boys to develop depression. According to Al-Amer, the difference in depression rates between adolescent girls and boys is due to girls' societal pressure about their gender identity. Other attributed factors include age, psychosocial context, family history, and political conflicts. The author provide recommendations in relation to dealing with this silent epidemic of depression among MENA adolescents. These recommendations include early detection and treatment, as well as the establishment of mental health services and effective interventions. In addition to these recommendations, the author highlight the importance of prevention and suggest that various strategies should target individuals at a young age in order to buffer the rise in depression among adolescents, particularly for those who have high familial risks, irritability, and fear or anxiety in childhood and those who live in areas of conflict. Prevention methods for depression will need to include family-based programs that integrate effective treatment of parental depression and irritability. The author suggest using the standard interviews based on the Structured Clinical Interview for DSM Disorders (SCID) (5) and the Composite International Diagnostic Interview (CIDI) to diagnose depression among adolescents.

In the following chapter, Niland conducts a literature review of qualitative studies on mental health stigma among youth in MENA. She presents results of the Arab Youth Survey (2019) that consisted of face-to-face interviews with 3300 Arab youth from 15 countries within MENA. Results suggested that 31% of participants know of someone with mental health issues. Importantly, while 49% of participants viewed mental illness as a normal part of life, 50% viewed it negatively. Niland reports findings of a longitudinal survey of over 15,000 young people (aged 10–29 years) from Egypt, which found increased psychiatric disturbance, nervousness, uncertainty and reduced self-worth due to lower school achievement, being out of work, failure to marry and find a job, and failure to finish school. Niland concludes that the rigid sociocultural beliefs create a public stigma for mental illness that, in turn, induce self-stigma feelings of low self-esteem, self-efficacy, and fear and shame for experiencing mental illness. Her review uncovered various factors that relate to adolescent mental illness in MENA; these include education, gender, and familiarity with mental illness, religious teachings, cultural stereotypes, and family traditions. Niland concludes that mental health stigma hinders intervention research in MENA, and suggests that effective interventions must engage Arab youth in their everyday context. The author notes that by learning more about the challenges and tensions that adolescents face, researchers and health professionals can try to address the issue from their perspectives, taking account of family and sociocultural context. The author suggests that MENA governments should create or adopt effective programs that target adolescent mental health. Similar programs include the *Youth Mental Health Project* – an online resource program that was set up to combat New Zealand's high youth mild to moderate mental health problems – and *New Zealand's Youthline 2017*, as well as programs similar to youth resilience sessions. The latter teach strategies and skills to cope with difficult and challenging times in youths' lives, and connect mentors (older students) to work with youth and their family to help them achieve their goals.

In relation to policy and programs that target adolescent mental health in MENA, Dabis and Allabadi report that only 9 out of the 22 countries in the Arab region have developed or are in the process of establishing youth policies. The authors outlined existing youth policies and related gaps that target the needs of youth living in Palestine, Lebanon, Egypt, and Tunisia. Dabis and Allabadi highlighted that the ongoing political conflict in Palestine results in challenges for youth and hinders them from fulfilling their real potential and becoming active citizens. These challenges include daily youth exposures to high rates of violence, discrimination, and movement restrictions. Furthermore, significant military presence and checkpoints cause significant stressors for Palestinian youth who constantly fear harassment, beatings, and shootings. The authors report that Palestinian youth, who compromise 30% of the total population, lack access to quality youth-friendly healthcare including sexual and reproductive health, as well as opportunities to participate in public and political life. In Lebanon, where individuals between 15 and 29 years old constitute 28% of the population, key challenges that Lebanese youth face are education disparities between private and public schools, lack of information and services on sexual reproductive health (SRH), and high unemployment levels reaching 66%. Reproductive health services regarding youth and adolescents are in dire need of improvement in Egypt, as reproductive health services are only available to married females. The authors recommend that local institutions, schools, and family centers should address harmful practices and should include adolescents. In Tunisia, where individuals between the ages of 15 and 19 years old constitute 29% of the total population and 43% of the working age group, social and cultural issues contribute to mental illness and are in dire need of addressing. For example, unemployment rates among youth increased to 32% after the revolution. Alarmingly, despite the legal ban, the Demographic and Health Surveys in 2014 revealed that 92% of females have had female genital mutilation.

Alongside these issues, Dabis and Allabadi note that youth concerns and policies have been on the Palestinian agenda during recent years, aiming to target the most vulnerable youth. Policy agendas include key focus areas such as economic empowerment that address the high unemployment, poor quality of education, and gender-normative labor segregation; civic engagement to offset poor youth representation and limited decision-making power; and access to health including holistic services, and sexual and reproductive health. The authors suggest that public authorities need to focus on national strategies to provide young people with opportunities and experiences that support their successful integration into society. This necessitates the involvement of youth councils, youth non-governmental organizations, interest groups, youth groups, youth workers, youth researchers, schools, employers, medical staff, faith groups, and the media in order to ensure a comprehensive participatory approach.

Adolescent health in the Kingdom of Saudi Arabia is the focus on the next chapter. Abdul Salam and Abdul Rashid note that transport injuries, unintentional injuries, and cardiovascular and circulatory diseases are the top three main causes of deaths for Saudi adolescents. Other health concerns include diarrheal diseases, lower respiratory illnesses, meningitis, cancer, non-transmissible diseases, malaria, diabetes, HIV, and tuberculosis. The authors state that these primary health issues contribute to depression and anxiety among adolescents. Specifically, the disability-adjusted life years (DALY) are disproportionately higher among adolescents in the Arab region, with mental and behavioral disorders as the primary cause of DALYs, followed by transport injuries. The authors present a governmental initiative, *Vision 2030*, which is a project that tackles the primary prevalent issues in the Saudi population. They note that the initiative highlights three main directive dimensions: awareness and knowledge of adolescents, effective listening and communication with adolescents, and delivering needed healthcare. Some of *Vision 2030* goals include primary prevention through the promotion of a healthful diet and active lifestyle from pre-school days, sexual awareness and education, engaging parents in childhood parenting programs, and encouragement of adolescent health research through developments such as health surveillance, nutritional database, and diagnostic tools for depression and anxiety. In addition, the initiative places greater emphasis on adolescents' challenges and concerns, support with social and digital media, and widened access to healthcare facilities and services with fully trained physicians.

In the last chapter, Alkhaldi et al. note several factors that contribute to physical and mental health issues among adolescents from MENA. The authors delineate these issues between occupied and politically unstable countries such as the Syrian Arab Republic, Lebanon, and Palestine and the economically advanced Arab Gulf countries. Within the former, issues include lack of education and health services for adolescent refugees and unsanitary conditions within refugee camps, which increase the risk of transmission of communicable diseases among adolescents. They note that evidence from the Arab Gulf countries suggests that adolescents demonstrate health-compromising behaviors and conditions, such as tobacco use, lack of physical activity, poor nutrition, and obesity. The authors conclude that current national health strategies in various MENA countries do not address adolescent health, and that many regions fail to monitor factors that influence the health status of adolescents. Thus, countries within MENA are in dire need of enhancing healthcare practice, healthcare facilities, clinical education, comprehensive school health services program, and adolescent health research to address key aspects of adolescent health and healthcare. Alkhaldi et al. highlight three main recommendations that focus on policy and intervention to improve adolescent health, particularly given our experience in the ongoing COVID-19 pandemic. The authors emphasize the need for educating adolescents and increasing awareness on global issues and challenges that assist in building resilience and facilitate good mental and physical health. They highlight the need for economic support and surveillance, especially for impoverished individuals who have been left jobless or whose jobs pose a grave risk for COVID-19 transmission. Finally, adolescent health in MENA can benefit from partnerships between non-governmental agencies and local governments that focus on refugee and internally displaced people (IDP) camps to implement culturally and religious-sensitive and effective adolescent health programs.

1.2 Future Areas of Research

While the information highlighted in this book is not comprehensive in relation to adolescent mental health in MENA, three main related points emerge and are essential pillars for reducing the burden of mental illness among MENA adolescents. First, it is evident that diverse social, cultural, environmental, and individual factors influence the mental health of adolescents from MENA. These factors tend to be unique to the region and compound background factors related to normal growth and development that adolescents experience in other countries. For example, while factors such as parenting or school achievements generally shape self-identity, motivation, and resilience of adolescents, those who live in certain MENA regions are also experiencing regional conflicts and wars that influence their self-identify, motivation, and resilience in different ways. No doubt that these additional factors are unique predictors of adolescent mental health in MENA. Furthermore, evidence is growing that the prevalence of mental health issues among adolescents from MENA is relatively high and continuously on the rise given the most recent global issues, including numerous regional conflicts in Syria, Iraq, Yemen, Lebanon, Palestine, and Egypt, as well as the COVID-19 pandemic. Therefore, a defined focus on adolescent mental health for MENA is not only necessary for the immediate future, but also a major determinant of the future of population health in the region for the coming generations.

Second, it is evident that stigma exists in relation to adolescent mental health in MENA. This further magnifies the situation given that most adolescents and their parents will refrain from seeking help or addressing mental health issues. Generally, cultural, religious, and social beliefs in MENA relate outcomes of mental health illnesses as taboos or signs of weaknesses. These include the need for treatment or medication, psychotherapy, and suicide or suicide attempts. Therefore, deep understanding of issues that link to adolescent mental health necessitates that stigma is primarily and effectively addressed, leading to its reduction. It is essential that adolescents do not experience self-blame or shame in relation to their mental health experiences. An acceptance and understanding of adolescent mental illnesses will facilitate related research and contribute to knowledge development and dissemination that has the capacity to reduce the burden of adolescent mental health in MENA.

Lastly, it appears that the majority of local and national youth policy initiatives and programs within MENA emphasize various important dimensions of adolescent health, such as non-communicable diseases and sexual and reproductive health. However, most policies fail at the level of implementation to target adolescent mental health issues explicitly, the six dimensions of wellness for adolescents, and surveillance of adolescent mental health illnesses such as depression and anxiety. Given the high prevalence rate of depression among adolescents from MENA, despite the possibility of underestimates given the stigma, adolescent mental health should be on the agenda of every agency and organization in all MENA nations. Not only would programs and policies on adolescent mental health have the capacity to improve the general health of adolescents for MENA, such programs and policies have the capacity to improve and sustain a high level of population health, which is an essential and unique national resource for all MENA countries.

References

Al Makadma A. (2017). Adolescent health and health care in the Arab Gulf countries: Today's needs and tomorrow's challenges. *International Journal of Pediatrics and Adolescent Medicine 4*, pp.1–8

AlBuhairan, Fadia (2015). Adolescent and Young Adult Health in the Arab Region: Where We Are and What We Must Do. *Journal of Adolescent Health.* Vol.57, Issue 3, pp.249-251. https://doi.org/10.1016/j.jadohealth.2015.06.010

Al-Krenawi, Alean. (2005). Mental health practice in Arab countries. *Curr Opin Psychiatry*, 18:560-564.

CDC. (2018, January 26). Learn About Mental Health - Mental Health - CDC. Retrieved January 25, 2021, from https://www.cdc.gov/mentalhealth/learn/index.htm

Dahlgren, G., Whitehead, M. (1991). Policies and strategies to promote equity in health. Stockholm: Institute for Future Studies.

Fuhrmann, D., Knoll, LJ., Blakemore, SJ. (2015). Adolescence as a sensitive period of brain development. *Trends in Cognitive Sciences, 19*(10), 558–566.

Galderisi, S., Heinz, A., Kastrup, M., Beezhold, J., & Sartorius, N. (2015). Toward a new definition of mental health. *World Psychiatry, 14*(2), 231.

Giedd, J.N., et al. (1999) Brain development during childhood and adolescence: a 222 longitudinal MRI study. Nat. Neurosci., 861–863.

Hoebeke, V. (2019, March 12). 4 Ways to Cultivate Intellectual Wellness. Retrieved February 02, 2021, from https://www.rtor.org/2015/08/11/4-ways-to-cultivate-intellectual-wellness/

Karamouzian, M., & Madani, N. (2020). COVID-19 response in the Middle East and north Africa: Challenges and paths forward. *The Lancet Global Health, 8*(7), e886–e887. https://doi.org/10.1016/S2214-109X(20)30233-3

Marks, D. F. (2002). Freedom, Responsibility and Power: Contrasting Approaches to Health Psychology. *Journal of Health Psychology 7*(1), 5–19.

National Alliance on Mental Illness. (2017). Mental Health Conditions. Retrieved January 26, 2021, from https://www.nami.org/learn-more/mental-health-conditions

Ostby, Y., et al. (2011) Morphometry and connectivity of the fronto-parietal verbal working memory network in development. *Neuropsychologia 49*, 3854–3862.

Piaget, J., et al. (1977) Epistemology and psychology of functions. (Trans J. Castellanos & V. Anderson). D. Reidel

Provencher, H. L., & Keyes, C. L. M. (2011). Complete mental health recovery: Bridging mental illness with positive mental health. *Journal of Public Mental Health, 10*(1), 57-69. http://dx.doi.org.zulib.idm.oclc.org/10.1108/17465721111134556

Ritchie, H., & Roser, M. (2018, January 20). Mental Health. Retrieved January 27, 2021, from https://ourworldindata.org/mental-health#:~:text=Globally%20an%20estimated%20284%20million,experience%20anxiety%20disorders%20than%20men.

Roddick, M. L. (2016, September 14). The 8 Dimensions of Wellness: Where Do You Fit In? Retrieved February 02, 2021, from https://www.goodtherapy.org/blog/8-dimensions-of-wellness-where-do-you-fit-in-0527164

Ronald, W., Carol, D., Elsie, J., Lela, R., Satvinder, D., & Tara, W. (2010). Evolving definitions of mental illness and wellness. *Preventing Chronic disease*, *7*(1), 2.

Sebastian, C.L., et al. (2012) Neural processing associated with cognitive and affective Theory of Mind in adolescents and adults. Soc. Cogn. Affec. Neurosci. 7, 53-63

Squeglia, L.M., et al. (2013) Early adolescent cortical thinning is related to better neuropsychological performance. J. Int. Neuropsychol. Soc. 19, 962-970 236 16

Tamnes, C.K. et al. (2010) Brain maturation in adolescence and young adulthood: regional age-related changes in cortical thickness and white matter volume and 209 microstructure. Cereb. Cortex 20, 534-548

United Nations Children's Fund (UNICEF) (2019). *MENA Generation 2030: Investing in children and youth today to secure a prosperous region*. Available at: tomorrowhttps://www.unicef.org/mena/media/4141/file/MENA-Gen2030.pdf (Accessed 2 June 2020).

WHO (2016). The Global AA-HA Framework (Accelerated Action for the Health of Adolescents). Towards implementation of the Global Strategy for Women's, Children's and Adolescents' Health (2016-2030).

WHO. (2019, November 28). Mental disorders. Retrieved January 27, 2021, from https://www.who.int/news-room/fact-sheets/detail/mental-disorders

World Atlas, https://www.worldatlas.com/articles/what-are-the-mena-countries.html

Part II
Mental Health in the MENA

There Is No Health Without Mental Health: The Middle East and North Africa

Justin Thomas and Alaa Galadari

1 Context: Happy to Unhappy Arabia?

Melancholy affects the soul through fear and sadness... Sadness is defined by the loss of what one loves; fear is the expectation of misfortune.

'Imran Ibn Ishaq 10th C.E

Parts of the Arab world, namely, Yemen and Oman, were referred to as Arabia Felix – happy Arabia by the ancient Romans. Actual levels of happiness in Arabia during the bygone era are wide open to historical speculation. The presence, however, of mental health issues has been clearly documented. One of the oldest manuscripts exploring depressive illness was penned by the Arab physician Ishaq Ibn Imran. His work, Maqaal 'ala malaakhoolia (Treatise on Melancholy), was reported to be published in Tunisia as early as the tenth century (Omrani et al. 2012). His manuscript removes any doubt that mental health issues such as depression and bipolar affective disorder were described and discussed in medieval Arabia. It illustrated that mental health issues were viewed by scholars as treatable mental health ailments rather than demonic possession, weakness in character, or irreligiosity.

The Arab world has been known to diagnose and treat mental health problems as early as the ninth century. Arab scholars from medieval Arabia similarly identified one of the earliest accounts of eating disorders such as anorexia nervosa and published a case study in ninth-century Syria. The case study carried much resemblance to contemporary diagnostic criteria of eating disorders, anorexia nervosa in

J. Thomas (✉)
Department of Natural Sciences and Public Health, Zayed University, Abu Dhabi, United Arab Emirates
e-mail: justin.thomas@zu.ac.ae

A. Galadari
Sheikh Khalifa Medical City, Abu Dhabi, United Arab Emirates

C. Barakat, F. Al Anouti (eds.), *Adolescent Mental Health in The Middle East and North Africa*, Global Perspectives on Health Geography,
https://doi.org/10.1007/978-3-030-91790-6_2

particular (Hajal 1982). It described an adolescent prince (al Mu'tazzbillah) who refuses food or nourishment without any known or identified physiological cause. Dr. Bukhtishu Ibn Jibrail treated the young prince by using an early form of behavioral psychotherapy to encourage him back to health. Similarly, Colman (1993) described tenth-century polymath Ibn Sina to have documented the earliest description of anorexia nervosa.

Today, the world is witnessing a steady rise in the prevalence of common mental health disorders. Over the past few decades, both depression and eating disorders have been on the rise among Arab youth (for reviews see Thomas 2014).

2 Method

This chapter identifies relevant regional literature published in English from 1970 to 2020 using main search engines (e.g., PsychLit, ScienceDirect, PubMed, PsycInfo). Our research focuses on MENA populations (especially the GCC), with a focus on depression and eating disorders. This literature is then synthesized into the narrative reviews that make up much the remainder of this chapter.

3 Results

3.1 Depression

Well-conducted epidemiological studies exploring mood disorders in the MENA region are relatively rare. Such studies, however, have reported depressive prevalence (lifetime) ranging between 13% and 18%. Depressive risk factors, regionally speaking, include things like stressful life events (including financial stressors), being female, poor social support, and chronic illnesses (Abdul Razzak et al. 2019).

In general, explorations of depressive symptoms routinely rely on self-report inventories such as the Beck Depression Inventory version 2 (BDI-II) (Beck et al. 1996). This 21-item self-report inventory has been widely used in adult and adolescent populations around the world. Easy to administer and score (Beck et al. 1988), respondents rate the severity of depressive symptoms experienced over the past 2 weeks, ranging from not all (0) to severe (3). Numerous studies have explored the inventory's psychometric properties, assessing how well it actually measures depression. Generally the inventory's items can be meaningfully categorized into two symptom groups: cognitive and somatic symptoms. Furthermore, there is fairly broad agreement that the BDI-II is a valid assessment of depressive symptoms across various cultures. For example, among 502 ethnically diverse college students in the USA (White 59%, Hispanic 22%, African-American 10%, Native American 10%, and Asian-American 7%), the BDI-II demonstrated similar psychometric properties for all groups, suggesting that the instrument is appropriate for use among populations of ethnically and culturally diverse college students. Additionally, studies using translated versions of the BDI-II, in nations as varied as the UAE, China, and Iran, generally evaluate the BDI-II as fit-for-purpose (depression screening and symptom assessment) within these internationally varied populations.

Measuring Depression in the Arab World

Numerous studies in the MENA region have also explored the psychometric properties (validity and reliability) of the BDI-II on Arab populations. One study looked at 9700 university students across 18 Arab nations. This study led by Alansari (2005) examined the psychometric properties of the Arabic BDI-II, reporting excellent content validity and internal reliability. The Arabic BDI-II also captured both cognitive and somatic/affective factors across this large international Arab population. The BDI-II appears to be a valid and reliable measure for use in the MENA region and cross-cultural contexts in general.

BDI Scores for Arab Youth

Over half the MENA region population (54%) are under the age of 25, so exploring depressive symptoms among youthful populations is particularly important in this context. The figure (Table 1) below details a sample of regional studies utilizing the BDI-II with Arab youth populations, mostly college students. These data are on par with similarly designed studies among college-age populations in North America and Europe. It is worth noting that the BDI-II was originally normed on Canadian college students, with a mean 12.11 (Beck et al. 1996). Studies among Arab college students have not reported lower means than these original norms, and frequently report means that are much higher.

3.2 Eating Disorders

Eating disorders were formerly thought to be afflictions bound up with the sociocultural norms of affluent and industrialized Western countries – so-called culturally bound syndromes. The cultural specificity of eating disorders like anorexia and bulimia begins to vanish with the rise of globalization and the internationalization of psychiatric research (Gordon 2000). Mervat Nasser, who researched disordered eating attitudes among female high school and university students, is a pioneer in the field. The research confirmed beyond a shadow of a doubt that non-Western populations also experienced disordered eating attitudes and eating disorders (Nasser 1986, 1992, 1994). One of her research studies compared the eating habits of female students at Cairo University to those of similar Egyptian female students studying

Table 1 Mean Beck Depression Inventory (BDI) scores for Arab university students

Study	State	Sample	BDI male M (SD)	BDI female M (SD)	BDI total M (SD)
Al-Musawi (2001)	Bahrain	200 (63% female) undergraduates at the University of Bahrain	13.39 (18.80)	13.49 (9.81)	13.44 (6.74)
Thomas et al. (2010b)	UAE	261 female students at Zayed University	Na	13.48 (7.21)	13.48 (7.21)
Baroun (2007)	Kuwait	292 undergraduate students at Kuwait University	18.26 (10.59)	15.68 (9.06)	Not reported
	UAE	450 (68.4% female students at Zayed University	13.67 (7.64)	14.47 (7.94)	14.22 (8.51)
	Several	University students aged 18–25 (50% female) in each state			Not reported
	Oman	236	14.3 (9.33)	18.3 [a](0.96)	
	Kuwait	610	13.0 (9.73)	15.0 [a](9.17)	
	Qatar	200	12.3 (8.24)	15.1 [a](9.05)	
	Saudi Arabia	630	18.1 (11.4)	14.2 [a](10.1)	
	UAE	150	19.3 (13.0)	17.8 (11.2)	
Shah et al. (2020)	UAE	518 adolescent students in Al Ain	28 (13.9)	61 (19.2)	20.9 (4.5)
Dardas et al. (2019)	Jordan	3292 adolescent students in Jordan			16.7 (12.3)
Dardas et al. (2018)	Jordan	2349 Jordanian adolescents	17.8 (6.3)	13.6 (5.2)	16.3 (11.2)

[a] Statistically significant gender difference ($p < 0.05$)

in the UK. Using the Eating Attitudes Test, Nasser examined both groups for eating disorders (EAT).When Nasser examined the two groups, she discovered that 22% of female students in the UK scored above the EAT cutoff, compared to only 12% of female students in Egypt.

A considerable majority of the UK-based group matched the entire diagnostic criteria for an eating disorder, aside from the shockingly high levels of disordered eating attitudes identified in the Egypt-based group, which was previously regarded to be low risk. Nasser arrived at the tentative conclusion that (1) disordered eating attitudes and eating disorders are more common in non-Western populations than previously thought, and (2) increased exposure to Western cultural norms is linked to an increased risk of developing abnormal eating attitudes and eating disorders.

While it is now clear that eating disorders are not culture-specific, it is also clear that they are pathoplastic, that is, shaped by sociocultural factors. Such factors can influence both the expression and prevalence of eating disorders. This means that the symptomology and epidemiology of disordered eating are likely to alter as societies changed. For this reason it is important to build bridges between sociopsychological concerns and biological ones, taking into consideration the "meaning" of anorexia and bulimia for the patient and members of their society.

Socioeconomic Perspective

The Gulf and Middle East Arab countries differ in their openness to Western culture and lifestyle and are marked by social transformation, acculturation, and rapid economic growth. This is especially true in the Gulf states. According to a World Health Organization assessment on the UAE, few countries on the planet have seen such dramatic and far-reaching change in recent decades. In contrast to the more "collectivist" tribal lifestyle in the Gulf states decades before, the total change in lifestyle refers to the movement to a more individualistic and high-consuming lifestyle. Over the last few decades, Gulf countries' diets have shifted away from whole grains, fruits, vegetables, and fibers and toward a fast-food diet high in fat, salt, and sugar. Dieting was unusual before to the oil boom because the food supply was limited. Corpulence was the standard of beauty since it was generally associated with prosperity, health, and reproduction (Alkazemi et al. 2018). The rise in the prevalence of chronic diseases such as diabetes, dyslipidemia, and hypertension was largely attributed to this considerable change in lifestyle (Eapen et al. 2006). Aside from the physical effects of the lifestyle change, psychological issues could occur or be worsened in such a situation. Furthermore, the coping techniques that people adopt or discover to help them cope with transition can have a negative influence on psychological well-being in and of itself.

Sociocultural Perspective

Eating disorders are explained from a sociocultural perspective as a dysfunctional reaction to societal changes that reflect physical changes in adolescents transitioning to adulthood.

The renowned anthropologist George Devereux eloquently asserts that certain disorders are understood as responses to unresolved tension and anxiety within a particular culture in his work on ethnic disorders and culturally reactive syndromes (Gordon 2000).

Over the years, researchers have looked into some of the aforementioned ethnic illnesses. Hysteria, for example, which was traditionally associated with an infantile hyper-femininity is now thought to have been influenced by the Victorian era's societal challenges and paradoxes. The focus on delicate femininity and ornamental sexuality in nineteenth-century Western civilization, set against a morally oppressive and hypocritical sexual code, is said to have given rise to hysteria and molded its often colorful manifestation (Gordon 2000; Smith-Rosenberg 1972). Eating disorders can be viewed in a similar sense, as they may develop as a result of contradictory cultural signals and messages encouraging overconsumption while concurrently promoting an almost unattainable slender body ideal for women, for example. The symptoms of culturally reactive illnesses are typically thought to be the magnifying and exaggeration of socially acceptable behaviors. Dieting, weight reduction, and the overvaluation of thinness are examples of eating disorders.

The Thin Body Cult in the Arabia

A thinner, more tubular, female body image ideal has taken hold in the MENA region, probably as a result of transition, globalization, and acculturation; rigorous diets and the celebration of extreme thinness are now commonly accepted. This has also occurred in the context of a consumer society, where obesity is sometimes referred to as an epidemic.

A simple trip to any supermarket will demonstrate the presence of the thin body obsession in the Arab world. Svelte starlets dominate the covers of magazines in Arabic and English, and these glossy publications ride the checkout conveyor with slim-tea and other magical weight-loss goods. Somewhat more subtly, a fashion doll named Fulla recently appeared in the region's toy stores. Fulla, a nubile Arab female, was introduced in 2003 by Dubai-based producer NewBoy and has nearly equal body size proportions to her North American counterpart Barbie. Fulla is actually a slightly slimmer and much more tubular than Barbie.

Fulla was created to reflect Muslim and Middle Eastern beliefs, according to the manufacturers, and she exhibits this element of her identity with her abaya, head scarf, and prayer mat. Fulla, however, indirectly represents the slender body cult, despite the evident trimmings of tradition. She is also the best-selling fashion doll in the Gulf region at the time of writing, with her owners claiming to have sold 1.5 million dolls, with 90% of sales originating from the Gulf (Harrod 2012).

3.3 *Eating Disorders in the Arab World*

According to the findings of the examined studies, rates of aberrant eating attitudes in the Middle East and North Africa are equivalent to those in Western nations. Five cases of anorexia nervosa were documented in the UAE and neighboring Oman, according to a study by UAE University and Al Ain's department of psychiatry (Abou-Saleh et al. 1996). Two of the patients were UAE nationals, two were Omanis, and one was Sudanese. There were three girls and two males among the patients. Both of the men had previously been obese and had been taunted about their weight when they were younger. Four of the five patients vividly reported a fear of fatness as a symptomatic of their disease, which was one of the most striking elements of the examined instances. This is crucial because a dread of fatness is regarded as a distinguishing element of current anorexia nervosa, although historical portrayals of the condition often lack this aspect (Bell 1985). Similarly, anorexia has been documented in several Asian societies, where dread of fatness is common and body image issues are nonexistent. Instead, physical factors such as bloating and stomach discomfort are commonly used to explain stringent dietary restriction. These more somatic patients adopt a different vocabulary of distress in the absence of body image concerns (fat-phobia), one centered on abdominal distension and discomfort rather than the abhorrence of obesity. A young Bedouin woman from the isolated rural area known as the Rub al-Khali was the only patient in the UAE case series who did not report body image problems or dread of fatness (the Empty Quarter). The lack of body image concern in this patient, according to the study's authors, may reflect the isolated Rub al-Khali people's less acculturated rural lifestyles.

The consequence is that in locations where globalization has had less of an impact, we witness a more traditional presentation of the condition, one that is free of body image issues. However, body image problems were present in four of the five cases discovered in this study, which the UAE research team believes is indicative of the UAE society's increasing "Westernization." Another regional case study, this time from Kuwait, focuses on a 14-year-old girl who displays all of the classic symptoms of anorexia nervosa (Qadan 2009). The patient had lost 15 kg in 6 months, her period had stopped 5 months before (secondary amenorrhea), and her BMI was 15. Excessive lanugo hair has also appeared on the patient's body. This is a unique form of hair growth that is linked to malnutrition. The patient revealed that her excessive dietary restrictions were motivated by an obsessive dread of gaining weight during clinical interviews. Despite her noticeable emaciation (BMI 15), she thought her weight and shape were perfect, and she was adamant about not gaining any weight.

Her doctor, who was also the author of the original case report, speculated that the condition may have developed as a result of new experiences such as changing schools and adapting into a thinness-conscious culture. A summer trip in Lebanon is also mentioned as a possible triggering factor, with the physician speculating that Lebanon's more "Westernized," body image and fashion-conscious society played a part. One of the major issues raised in this case was the complete lack of specialized care in the Gulf region to address similar illnesses. Following the discovery of unequivocal cases of anorexia nervosa in the Gulf in the 1990s (Abou-Saleh et al. 1996), various studies have sought to quantify the overall frequency of eating disorders, as well as to identify risk and resilience factors. These researches generally investigated the following questions: how widespread are eating disorders and eating disorder-related psychopathology in the Arabian Gulf, and what makes some people more or less vulnerable?

In a study conducted in Al Ain, UAE, researchers used a multiphase screening approach to look at eating disorder-related psychopathology among 495 female high school students (Eapen et al. 2006). During phase one of the trial, all students aged 13–18 years old completed the Eating Attitudes Test, a commonly used screening instrument. This instrument was created as a screening tool to detect those who are at risk of developing eating disorders, particularly anorexia. A score of more than 30 on this test indicates that you are at risk of developing an eating disorder.

High test results are usually followed by an in-depth clinical interview to confirm or rule out an eating disorder diagnosis. During phase one, students performed the EAT as well as additional assessments, such as a body size preference assessment, a drive for thinness evaluation, and questions on TV viewing habits. In all, 116 (23.4%) of the students exceeded the EAT cutoff level. In phase two of the study, a psychiatrist examined a random sample of 100 students, 50 of whom scored over the cutoff and 50 of whom scored below the cutoff, utilizing a standardized, structured clinical interview. 78% of those in the high-risk category (those with a score of more than 30) were classified as body-dissatisfied, indicating a wish to be slimmer, and 25 percent were diagnosed with subclinical anorexia nervosa, which is defined by an excessive dread of becoming fat and gaining weight. Two of the girls in the study truly satisfied all of the diagnostic criteria for anorexia nervosa and bulimia.

In addition, the study found a link between EAT scores and the desire to lose weight. "Knowing someone on a diet," "having a family member with a weight problem," and "watching Western TV programs" were all found to predict higher EAT scores. All of the statistics and associations found in this study are nearly identical to those found in earlier research on Western school-aged populations. The authors of the study urged for frequent screening for eating disorders as part of the UAE's school health program, citing the prevalence of body image dissatisfaction, dieting behavior, and inappropriate eating attitudes.

Another UAE study looked at female university students at an institution with campuses in Dubai and Abu Dhabi. The findings were very comparable, despite the fact that these women were substantially older than the Al Ain study's high school girls. Again, the frequency of eating disorders symptomology was equivalent to that found in research of college girls conducted in Western industrialized countries, and there were obvious links between body image dissatisfaction, acculturation indices, and eating disorder symptoms (Thomas et al. 2010).

Other Gulf countries have reported similar findings. The validity of an Arabic version of the Eating Attitudes Test for use as a screening tool in nonclinical groups was investigated by Al-Subaie et al. (1996). A total of 129 female high school students took the test, with 19.7% scoring higher than the instrument's cutoff. One of the students was diagnosed with anorexia nervosa after further diagnostic testing. The authors of the study suggested that the Arabic EAT could be beneficial for screening huge Saudi populations. Al-Subaie (2000) investigated the incidence of dieting and eating disorder-related behaviors among 1271 Saudi female high school students in another Saudi study. This study also tried to figure out how much specific sociocultural elements influenced these behaviors. The eating disorder inventory (EDI) was employed in the study, as well as information on acculturation. Participants were asked, for example, if they had spent more than 6 months in a Western country and if they spoke any Western languages. Several demographic variables, such as family size, parental education, and parental marital status, were also explored, as well as polygamy.

The study discovered that 15% of the subjects scored higher than the eating disorder inventory cutoff, a number comparable to research conducted in Western industrialized countries. More notably, having lived in a Western country for more than 6 months, speaking a Western language, being overweight (higher BMI), having a smaller family size, and having greater levels of parental education were the factors most predictive of eating disorder symptoms. With the exception of BMI, all of these variables can be interpreted as markers of acculturation to Western culture. Because many highly educated Saudis will have had at least some of their schooling in the West, parental education was judged an essential sign of acculturation (Al-Subaie 2000). In a survey of 144 Saudi women who visited a basic healthcare clinic in Al-Khobar, a big city in Saudi Arabia's eastern province, education was found to be an essential determinant. The study looked at the variations in body size estimates and body image ideals between normal and overweight women. One of the most important findings from this study was that there was a definite link between education and a desire for a thinner ideal body image.

Participants had been shown series of illustrations of a lady of various sizes, ranging from hyper-obese to extremely underweight, to represent a continuum of weight categories. The individuals with the highest levels of education tended to choose thinner models to symbolize their ideal body size and shape. According to the authors, educated Saudi women have more access to Western literature and are hence more vulnerable to developing Westernized body image norms. They claimed that uneducated women had more conventional Arabian ideas, in which beauty and health are equated with a "little fat" (Al-Subaie 2000).

However, access to modernity's global culture through visual media such as television could be a far more potent shaper of body image ideals than "Western literature."

The majority of regional study has focused solely on female citizens, ignoring comparisons with Western expatriate communities and males. One research in Oman included both males and females, as well as expats and locals (Al-Adawi et al. 2002). This study compared Omani teens (mean age 15.12, SD = 0.58) to Omani adults (mean age 38.71, SD = 5.43), taking into account both a cross-cultural and a cross-generational perspective.

In each group, there were 100 participants: 100 Omani adolescents, 100 Omani adults, and 100 Western expatriate adolescents. Males and females were about equal in number. The EAT was completed by all participants, as well as the BITE (Bulimic Investigatory Test Edinburgh), which is a specialized assessment of bulimic symptoms, whereas the EAT is more focused on anorectic-type behaviors. The findings were similar to those of previous Gulf research. On the EAT, nearly one-third of Omani adolescents exceeded the screening cutoff, but only 12% did so on the BITE. However, the fact that almost no Omani adults scored above the cutoff was even more intriguing (2%). This was a significant and statistically significant generational gap. Omani teenagers even had slightly greater levels of eating psychopathology than their age-matched Western expatriate peers.

The authors of the study use acculturation terminology to explain their findings. Younger Omanis appear to be significantly more influenced by the increasingly prevalent global culture of modernity, sometimes known as Westernization, as evidenced by the generational differences. Omani teenagers are becoming more fluent in English, and many of them are taught by foreign expatriate teachers. In other words, kids are gaining access to Western-influenced educational options that their parents have never had. Furthermore, due to the widespread use of global mass media, younger Omanis are exposed to cultural influences in the form of entertainment that would have been unthinkable to their parents during their formative years (Al-Adawi et al. 2002).

One writer eloquently expresses this notion of a quickly changing physical and psychological environment, stating, "Omani society, driven by oil funds, achieved in 20 years the kind of advancement that took Europe a thousand years" (Smith 1988 p.297). Other critics on Oman's development bemoan elements of the country's quick development, claiming that it has wreaked havoc on interpersonal connections and given rise to a society where individuality is increasingly acceptable and education has largely replaced family values (Melikian 1988).

3.4 Rethinking Eating Disorders

The concept of "Westernization," on the other hand, is a pretty broad and nebulous one. The word frequently connotes the oversimplified notion that eating disorders are disseminated by a mindless mimic attitude toward all things Western, particularly contemporary beauty and body image norms. This idea is appealing because of its simplicity; yet, the research data reveal a slightly more complicated picture. The indications of eating disorders among schoolgirls of Asian descent were investigated in a UK study conducted in the city of Bradford. Increased levels of eating disorder symptoms were linked to traditional cultural orientations rather than to Western cultural orientations in this study.

According to the authors, cultural adjustment stress, rather than just the effect of Western beauty norms, may play a role in the development of disordered eating (Mumford and Whitehouse 1994). Numerous other researches have also found that ethnic minority and immigrant communities in Western countries had a higher prevalence of eating disorder-like behaviors and symptoms (Robinson et al. 1996). These findings have led to the hypothesis that, in addition to aesthetic concerns, ethnic minorities' eating disorders may be caused by disputes over cultural and "racial identity."

Nasser (2009) underlines the need to look beyond the "Westernization" and explore the mechanisms that underpin it, such as consumerism, individualism, shifting gender roles, and increasing levels of alienation, in a review paper on culture and eating disorders. Nasser claims that "[individuals]...are obliged to resort to the 'body' as a new medium for expressing this cultural disorientation and distress" in such a scenario (p.347). Eating disorders, according to Nasser, are a "marker of transition," diagnostic of civilizations in transition, when cultural norms may have become compromised.

Eating disorders, according to Nasser, are never simply the result of being exposed to another culture's norms or of adopting Western body image ideals.

Regional demography may be playing a role in addition to the complicated multidimensional factors of acculturation and the strains of transition. The Gulf region has a younger population than developed Western nations. According to estimates a third of the population in each of the six GCC countries is under the age of 25 (AlMunajjed and Sabbagh 2011). It is suggested that such demographics, skewed toward youth, put additional strain on young people. For example, in such a demographic setting, young people face increased competition for social and vocational positions and may have to deal with the stigma and pain associated with failure more frequently (Thomas 2018). Individuals are also thought to be at a higher risk of suffering "adjustment difficulties" as a result of such demographic patterns. This clinical phrase denotes a person's inability to cope with a situational stressor (transition). Eating disorders are frequently considered as developing in such a setting, possibly as a maladaptive coping mechanism (Kleinman and Cohen 1997).

Another issue worth considering, especially in explaining generational disparities, is the concept of evolving distress idioms. This viewpoint meshes well with the idea of eating disorders being culturally reactive. Eating psychopathology has become a known and acceptable expression of distress within this younger population, which could explain the higher frequency of eating disordered symptoms in Omani adolescents compared to the previous generation. This concept is succinctly described by Al-Adawi et al. as "old distress in new bottles." Possibly, rather than somatic distress like their parents did, this younger generation is more "psychologically minded," meaning they are more likely to experience and express suffering in psychological terms rather with the somatic idioms of old, such as abdominal pain and migraines.

4 Discussion

Eating disorder symptomology is comparable to, if not greater than, research in Western industrialized nations, according to the regional studies evaluated. The same test, the Eating Attitudes Test, was used to assess them all (EAT). It was designed as a screening tool for anorexia nervosa, with scores above 30 (the cutoff) indicating the need for further study. A structured clinical interview with a psychiatrist or psychologist is usually the next step. It's vital to note that EAT scores aren't diagnostic of eating disorders on their own. A high EAT score may only be interpreted as indicating considerable levels of aberrant eating attitudes and behaviors; a score above the cutoff point necessitates a more thorough diagnostic examination. However, because the EAT is widely used, we can examine the overall pattern of findings across investigations. Looking at all of the research that used the EAT-26 and EAT-40, we can observe that between 19.6 and 44.5 percent of subjects scored above the cutoff. Similarly, other research that used a variety of eating and body image-related evaluations found high levels of dissatisfaction and distress, as well as eating disorder-associated behavior.

5 Conclusions

Mental health problems among adolescents, specifically eating disorders and major depressive disorder, are a public health concern for the MENA region. Given how such disorders contribute to disability, this relatively youthful population suggests an even greater need to respond to such issues. The prevalence of depression in the region has been found to range between 13% and 18%. Both depression and anorexia once established can become chronic, leading to years of distress and disability for the sufferer. Mental healthcare systems until recently have focused on treatment methods and how best to alleviate symptoms. While reducing symptoms does impact functioning and well-being, a more holistic approach ensures all aspects of a patient are considered – a concept that is central to mental health professionals. Understanding the complex relationship between mental health conditions and culture could change the way we deal and interact with conditions. Appreciating the interaction between biological and social risk factors may strengthen the understanding of mental health problems. But as more questions are answered, others arise. Have transcultural comparisons in psychopathology shrunk because of technology and mass communication? Has mass media become so widespread that people escape their lives using the same screens, creating a medium for cultural assimilation? The modern age brought with it capitalism, urbanization, information technology, advertisements, and the changing of social roles of women. We advocate prevention. Initiatives aimed at fostering resilience and emotional self-management, such as mindfulness-based stress reduction, should become part of educational curricular. To stem the tide of mental health problems among Arab youth, prevention is key.

Critical Thinking/Reflection Questions

1. Why are the rates of depression among the region's young people relatively high?
2. How might the region's religious traditions act as protective and risk factors in the context of mood and eating disorders and mental health in general?
3. In parts of the MENA region, there is a rising age for marriage; what impact might that have on mood and eating disorders?
4. In parts of the MENA region, consanguineous marriage is relatively common; what impact, if any, might this have on mood and eating disorders?
5. What is the best overall strategy to reduce the regional burden of mood and eating disorders and mental health problems in general?

References

Abdul Razzak, H., Harbi, A. and Ahli, S., 2019. Depression: Prevalence and Associated Risk Factors in the United Arab Emirates. Oman Medical Journal, 34(4), pp.274-282.

Abou-Saleh, M. T., Younis, Y. & Karim, L. 1996. Anorexia Nervosa in an Arab Culture. *International Journal of Eating Disorders,* 23, 207-12.

Al-Adawi, S., Dorvlo, A., Burke, D. T., Al-Bahlani, S., Martin, R. G. & Al-Ismaily, S. 2002. Presence and Severity of Anorexia and Bulimia Among Male and Female Omani and Non-Omani Adolescents. *Journal of the American Academy of Child & Adolescent Psychiatry,* 41, 1124-1130.

Alkazemi, D., Zafar, T.A., Ebrahim, M. and Kubow, S. (2018). Distorted weight perception correlates with disordered eating attitudes in Kuwaiti college women. *International Journal of Eating Disorders*, 51(5), pp.449–458.

Al-Musawi, N. M. 2001. Psychometric Properties of the Beck Depression Inventory–II With University Students in Bahrain. *Journal of Personality Assessment,* 77, 568-579.

Al-Subaie, A. S. 2000. Some Correlates of Dieting Behaviour in Saudi Schoolgirls. *International Journal of Eating Disorders,* 28, 242-46.

Al-Subaie, A. S., Al-Shammari, S. A., Bamgboye, E., Al-Sabhan, Al-Shehri, K. & Bannah, A. R. 1996. Validity of the Arabic Version of the Eating Attitude Test. *International Journal of Eating Disorders,* 20, 321-24.

Alansari, B. 2005. Beck Depression Inventory (BDI-II) Items Characteristics Among Undergraduate Students of Nineteen Islamic Countries. . *Social and Personality Psychology,* 33, 675-684.

Almunajjed, M. & Sabbagh, K. 2011. Youth in GCC Countries: Meeting the Challenge. *In:* COMPANY, B. (ed.). Abu Dhabi: Booz & Company.

Baroun, K. 2007. The impact of gender, anxiety, and depression on performance on pursuit rotor. *SOCIAL BEHAVIOR AND PERSONALITY, 2007, 35(5), 615-626,* 35, 615-626.

Beck, A. T., Steer, R. A. & Brown, G. 1996. *BDI-II Manual,* San Antonio, Harcourt Brace & Company.

Beck, A. T., Steer, R. A. & Carbin, M. G. 1988. Psychometric properties of the Beck Depression Inventory: Twenty-five years of evaluation. *Clinical Psychology Review,* 8, 77-100.

Bell, R. M. (1985). Holy Anorexia. University of Chicago Press.

Colman, A. M. 1993. *Facts , Fallacies and Frauds in Psychology,* London, Routledge.

Dardas, L.A., Shoqirat, N., Xu, H., Al-Khayat, A., Bani Ata, S., Shawashreh, A. and Simmons, L.A. (2019). Comparison of the performance of the Beck Depression Inventory-II and the Center for Epidemiologic Studies-Depression Scale in Arab adolescents. *Public Health Nursing*, 36:564–574.

Dardas, L.A., Silva, S.G., Smoski, M.J., Noonan, D. and Simmons, L.A. (2018). The prevalence of depressive symptoms among Arab adolescents: Findings from Jordan. *Public Health Nursing*, 35(2), pp.100–108.

Eapen, V., Mabrouk, A. A. & Bin-Othman, S. 2006. Disordered eating attitudes and symptomatology among adolescent girls in the United Arab Emirates. *Eating Behaviors*, 53-60.

Gordon, R. A. 2000. *Eating Disorders: Anatomy of a Social Epidemic,* Oxford, Blackwell.

Hajal, F. 1982. Psychological Treatment of Anorexia: A case from the Ninth Century. *Journal of the History of Medicine and Allied Sciences,* XXXVII, 325-328.

Harrod, H. 2012. Little girls, big business. *The Telegraph*, 12/June/2012.

Kleinman, A. & Cohen, A. 1997. Psychiatry's global challenge. *Scientific American,* 276, 86-89.

Melikian, L. 1988. Arab socio-political impact on Gulf life styles. *In:* PRIDHAM, B. R. (ed.) *The Arab Gulf and the Arab World.* London: Croom Helm.

Mumford, D. B. & Whitehouse, A. M. 1994. Sociocultural correlates of eating disorders among Asian school girls in Bradford. *British Journal Of Psychiatry,* 34, 677-88.

Nasser, M. 1986. Comparative study of the prevalence of abnormal eating attitudes among Arab female students at both London and Cairo universities. *Psychological Medicine,* 16, 621-25.

Nasser, M. 1992. *Screening for abnormal eating attitudes in a population of Egyptian secondary school girls.* Doctor of Medicine (DM), Southampton University.

Nasser, M. 1994. Screening for abnormal eating attitudes in a population of Egyptian secondary school girls. *Social Psychiatry and Psychiatric Epidemiology,* 29, 88-94.

Nasser, M. 2009. Eating disorders across cultures. *Psychiatry,* 8, 347-50.

Omrani, A., Holtzman, N. S., Akiskal , H. S. & Ghaemi, N. 2012. Ibn Imran's 10th century Treatise on Melancholy. *Journal of Affective Disorders,* 141, 116-119.

Qadan, L. 2009. Anorexia Nervosa: Beyond Boundaries. *International Journal of Eating Disorders,* 42, 479-81.

Robinson, T. N., Killen, J. D., Litt, I. F., Hammer, L. D., Wilson, D. M., Haydel, K. F., Hayward, C. & Taylor, C. B. 1996. Ethnicity and body dissatisfaction: Are Hispanic and Asian girls at increased risk for eating disorders? *Journal of Adolescent Health,* 19, 384-393.

Shah, S., Al Dhaheri, F., Albanna, A., Al Jaberi, N., Al Eissaee, S., Alshehhi, N., Al Shamisi, S., Al Hamez, M., Abdelrazeq, S., Grivna, M. and Betancourt, T., 2020. Self-esteem and other risk factors for depressive symptoms among adolescents in United Arab Emirates. PLOS ONE, 15(1), p.e0227483.

Smith, R. 1988. Oman: leaping across the centuries. *British Medical Journal,* 297, 540-544.

Smith-Rosenberg, C. 1972. The Hysterical Woman: Sex Roles and the Role Conflict in Nineteenth Century America. *Social Research,* 39, 652-78.

Thomas, J. 2014. Psychological Well-Being in the Gulf States: The New Arabia Felix, London, Palgrave Macmillan.

Thomas, J., Abdulrahman, A. & Khan, S. 2010. Eating attitudes and body image concerns among female university students in the United Arab Emirates. Appetite, 54, 595-598.

Thomas, J., Cambell, C., Altareb, B. & Ali, A. 2010b. Emotional Stroop interference for depression-related stimuli in a United Arab Emirates student population. Social Behaviour and Personality, 5, 597-600.

A Silent Epidemic of Depression Among Adolescents in the Middle East and North Africa Region: Emerging Tribulation

Rasmieh Al-amer

1 Introduction

Unipolar depressive disorder, otherwise known as depression or clinical depression, is a debilitating mood disorder common among adolescents worldwide (Thapar et al. 2012) and is the second leading cause of death in this age group. In particular, adolescents are identified as those who fall between 10 and 19 years of age (Csikszentmihalyi 2020). Over the past decade, incidences of depression have shown an upward trend across all age groups, with a noticeable increase among adolescents, reaching unprecedented levels worldwide (Thapar et al. 2012). To illustrate, the incidence of depression has increased from 172 million in 1990 to 258 million in 2017, indicating a 49.86% increase (Liu et al. 2020). This suggests that depression remains a pressing public health issue globally.

Today, around a quarter of adolescents are anticipated to suffer from at least one episode of severe depression by the age of 18, while around one tenth will have experienced at least one major depressive episode by the age of 14 (National Institute of Mental Health). Moreover, the WHO in 2005 reported that only a quarter of adolescents with depression seek treatment, one third of whom adhere to their treatment plan (World Health Organization 2005). Regrettably, adolescents' attitude toward their disease is mainly attributed to the stigmatization and stereotypes linked to mental illnesses, which add to the global disability and, hence, increase disease burden (Lopez et al. 2006). Moreover, adolescents with depression who are not on

R. Al-amer (✉)
School of Nursing, Isra University, Amman, Jordan

School of Nursing and Midwifery, Western Sydney University,
Penrith South, NSW, Australia
e-mail: R.al-amer@outlook.com

C. Barakat, F. Al Anouti (eds.), *Adolescent Mental Health in The Middle East and North Africa*, Global Perspectives on Health Geography,
https://doi.org/10.1007/978-3-030-91790-6_3

a treatment plan are at higher risk for multiple forbidding consequences, such as but not limited to suicide (Horesh et al. 2003) and health destructive behaviors (Dardas 2017). Of concern, depressive symptoms during the adolescence stage often persist or recur and continue into adulthood, indicating that depression during youth may predict more severe illnesses during adulthood (Lewinsohn et al. 2000).

2 The Middle East and North Africa (Mena) Region

Depression and its comorbidities have become major contributing factors to the global burden of chronic diseases in the Middle East and North Africa (MENA) region, with projections showing that depression will have become the second leading cause of disability by the year 2030. Furthermore, studies have shown that depression is expected to emerge as a major cause of disease burden in the future, mostly in developing countries (World Health Organization 2005). Unfortunately, adolescents from the MENA region have a greater proportion of disability-adjusted life years lost to depression compared to those in other regions, with females being more affected than their male counterparts (Obermeyer et al. 2015). This could be due to a vicious cycle of poverty and gender inequality (Khouri 2019). However, research has focused almost exclusively on high-income countries (Thapar et al. 2012). Although research may be a promising approach toward addressing depression among adolescents through the development of new strategies or improvement of existing services, data related to depression throughout the MENA region have been limited.

The MENA region constitutes of a high proportion of adolescents. In fact, a report from the United Nations stated that nearly one in five individuals in the MENA region is between 15 and 24 years old and that 50% are under 25 years of age (UNICEF 2011). Countries within the MENA region are classified as collectivist communities in which family members are obliged to help one another when health problems occur, with families occasionally seeking help from religious figures when mental health problems are encountered (Nydell 2005). This could be attributed to the lack of mental health professionals, limited resources, poor accessibility to health-care services, and most importantly stigma that precludes individuals from seeking professional help.

Moreover, the culture within the MENA region embraces superstitious beliefs (bad eyes, Hassad, and bad omens), which would lead them to view a mental health problem as an external unmodifiable issue rather than an internal, manageable disease (Heckman 2018). Furthermore, when Arabic individuals suffer from depression, they tend to communicate somatic symptoms and intentionally overlook emotional symptoms (Bener et al. 2013). However, the religious teachings of Islam and Christianity, the two main religious beliefs within the MENA region, clearly require and implore individuals to take care of their physical and mental health.

The current chapter discusses depression as a complex critical mental health problem among adolescents from the MENA region in terms of its proportion, risk factors (gender, age, psychosocial context, family history, political turmoil), and consequences, followed by some conclusions and recommendations.

3 Research Methodology

The author carried out a systematic search of the PubMed and CINAHL databases, both of which were selected given their comprehensive abstracts and indexing of most of the published literature in health-related fields. Moreover, CINAHL has been considered one of the largest databases for nursing literature in particular. The following search terms were used (depression, depression disorders, depressive, depressive disorders, adolescent, teen, youth, MENA region, Middle East). The search has identified 5025 citations from literature search of PubMed and CINAHL electronic databases (duplicates removed), and 4960 citations excluded based on screening titles and/or abstracts for inclusion and exclusion criteria with 23 studies included in the review (Fig. 1).

3.1 Characteristics of the Included Studies

All studies were selected based on the following inclusion criteria: reported on adolescents that live within the MENA region; reported on the prevalence rate of depression; used standardized, valid, and reliable scales; and followed a cross-sectional design. The inclusion criteria were tabulated on an Excel sheet and the author checked each paper against the mentioned criteria. Afterward, the author asked a colleague "Dr. M.D." to double-check the papers against the same inclusion criteria independently. Once we finalized the included papers, the author started the data extraction in which she has extracted the following data: the prevalence of depression among the study participants, risk factors for depression, the influence of gender, age, psychosocial context, family history, political climate, and consequences of depression. Additionally, the same colleague (Dr. M.D) was asked to check the data extraction independently, and any discrepancies were resolved by discussion.

4 Findings

The following section will discuss the prevalence rate of depression among adolescents from the MENA region, as well as its risk factors, consequences, and recommendations.

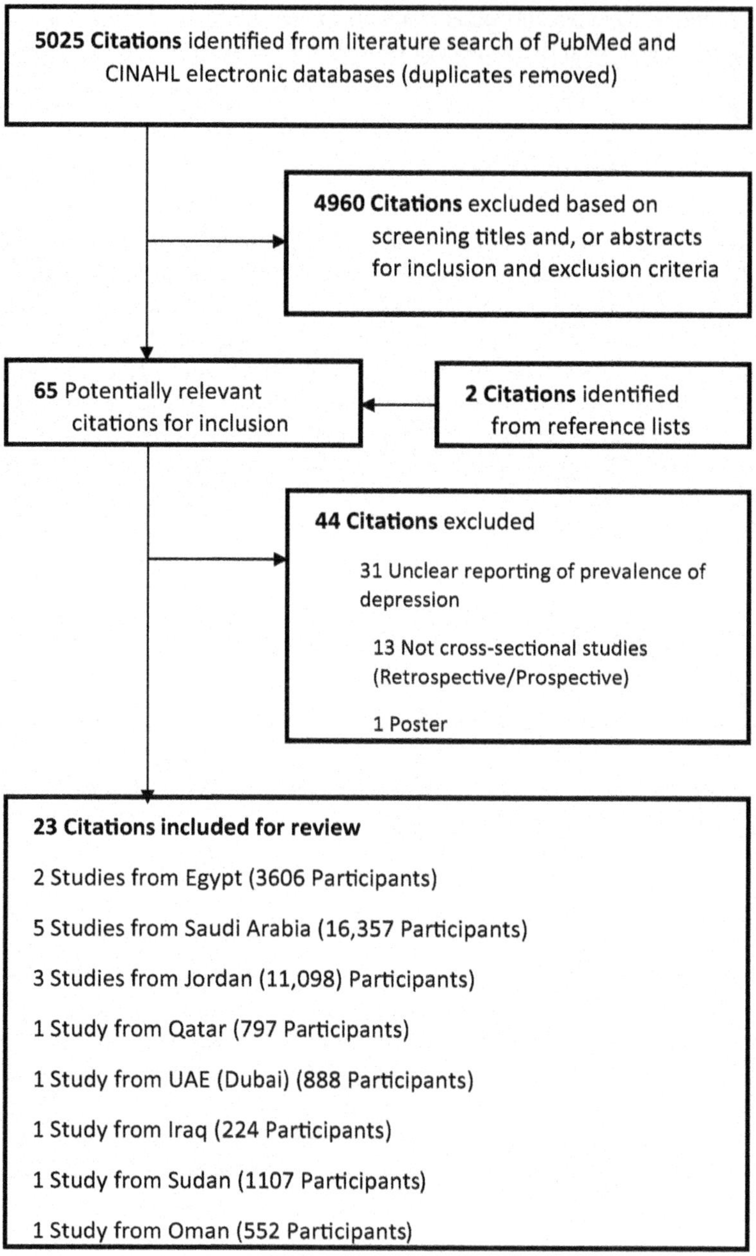

Fig. 1 Flowchart summarizing identification of studies included for review

4.1 *Prevalence Rate of Depression Among Adolescents from the MENA Region*

Epidemiologic studies that have investigated the prevalence of depression among adolescents from the MENA region have yielded inconsistent results (Table 1). This section summarizes the results of 23 studies on adolescents, among which 2 were from Egypt, 5 from Saudi Arabia, and the remainder were based in Jordan ($n = 3$), Palestine – Gaza ($n = 3$), Israel ($n = 1$), Kuwait ($n = 2$), Iran ($n = 2$), Qatar ($n = 1$), Iraq ($n = 1$), United Arab Emirates – Dubai ($n = 1$), Sudan ($n = 1$), and Oman ($n = 1$). Table 1 provides an overview of the epidemiologic studies that have been included in this chapter.

Overall prevalence rates of depression ranged from 11% in Sudan (Abdullatif 1995) to 41.2% in Palestine (Gaza) (Elbedour et al. 2007). Reported prevalence rates of depression among adolescents were between 13.9% and 41.5% in Saudi Arabia (Al Gelban 2009; AlBuhairan et al. 2015; Desouky et al. 2015; Asal and Abdel-Fattah 2007; Al-Gelban 2007); 17% in Egypt (Afifi 2007), 30% in Qatar (Al-Kaabi et al. 2017), 17.5% in Dubai (Ali et al. 2014), 11.8% in Kuwait (Abdullatif 1995), between 40% and 43% in Jordan (Al-Amer et al. 2019; Dardas 2017; Ismayilova et al. 2013), between 43.55% and 13.5% in Iran (Modabernia et al. 2007), 41% in Palestine (Gaza) (Elbedour et al. 2007; Kolltveit et al. 2012; Hein et al. 2007), 63.4% in Iraq (Freh 2016), 11% in females from Sudan (Shaaban and Baashar 2003), 3.7% in males and 4.8% in females from Kuwait (Abdullatif 1995), and 32% in Israel (Horesh et al. 2003).

Quite noticeably, the collectivist nature of these communities precluded researchers from precisely assessing depression among adolescents. To illustrate, adolescents are bound by the norms and culture of their community, which prevent them from providing data related to mental health problems, perhaps due to fear of social stigma. In fact, a person with depression would have been excluded from the community and would have had limited opportunities for social relationships, marriage, and employment (El-Islam 1982). Furthermore, depression usually coexists with anxiety and stress or other mental health problems, which could complicate the diagnostic process among adolescents with depression (Asal and Abdel-Fattah 2007).

Limited data have been available on the prevalence of depression among adolescents from the MENA region given the lack of population-based clinical screening. All 22 studies included herein used self-reporting scales, which have been known to result in higher proportions of depression compared to clinical interviews considering that they depend on symptoms rather than psychiatric diagnosis (Desouky et al. 2015). Hence, establishing prevalence rates of depression from the MENA region based on the International Classification of Diseases, Tenth Revision (ICD-10) or Diagnostic and Statistical Manual of Mental Disorders-5 (DSM-5) clinical interviews is of considerable importance.

Based on the aforementioned facts, our results suggest that depression rates among adolescents from the MENA region exceeded those reported by several international studies (Al Gelban 2009). In Europe, the prevalence rates of

Table 1 Characteristics and an overview of the epidemiologic studies that have investigated the prevalence of depression among adolescents from the MENA region

	Study	Data	Location	Findings
1	Ghareeb and Beshai (1989)	2029 adolescent students Age: 7–16 years Arabic version of the Children's Depression Inventory (CDI)	Egypt	Depression scores were below the cutoff for depression screening but higher than that reported in Western literature
2	Afifi (2007)	School adolescents 1577. Age: 14–19 Children's Depression Inventory	Egypt	Prevalence rate of depression 17.0%
3	Afifi (2006)	552 school adolescents Age: 14–20 Children's Depression Inventory	Oman	Boys (10.6%) were more likely to be depressed than girls (7.7%)
4	Al-Kaabi et al. (2017)	797 school adolescents Age group: 14–19 Arabic version of Beck's Depression Inventory-II (BDI-II)	Qatar	Prevalence rate of depression 34.5%
5	Al-Gelban (2007)	1723 male students particpated in this study Age: 15–19 Depression, Anxiety and Stress Scale (DASS)	Saudi	Prevalence rate of depression 38.2%
6	Asal and Abdel-Fattah (2007)	490 school adolescents Beck's Depression Inventory (CBDI) Age: 16–20 (BDI) cutoff 19 Arabic version	Saudi	Prevalence rate of depression 34%
7	Al Gelban (2009)	545 school adolescent (females) checklist 90 (SCL 90-R) Age group: 14–19	Saudi	The prevalence of symptoms of depression 13.9%
8	AlBuhairan et al. (2015)	12,575 school adolescents Age: 15.8 mean age Self-reported mental health symptoms	Saudi	Prevalence rate of depression 14%
9	Desouky et al. (2015)	1024 secondary school female students. BDI-24 Age: 15–17	Saudi Arabia	Prevalence rate of depression 42.9%
10	Ismayilova et al. (2013)	(*N* = 8129) Jordanian adolescents. Age: 14–25 Depression was measured on a nine-item scale	Jordan	Participants experiencing feelings of sadness (66%), loss of joy (49%), loss of hope in living (43%)

(continued)

Table 1 (continued)

	Study	Data	Location	Findings
11	Dardas et al. (2017)	2349 school adolescents Beck's Depression Inventory-II Age:12–17	Jordan	Prevalence rate of depression 41% of girls and 26% of boys
12	Al-Amer et al. (2019)	620 Palestinian refugees' adolescents A nine-item Patient Health Questionnaire for Adolescents (PHQ-A) Age: 13–18	Jordan	Prevalence rate of depression 40% Risk factor: female gender Recommendation: early screening
13	Ali et al. (2014)	888 school adolescents Age: 14–15 Children's Depression Inventory (CDI)	Dubai	Prevalence rate of depression 17.5%
14	Freh (2016)	224 adolescents living in Iraq Age:12–23 Beck's Depression Inventory (BDI-21)	Iraq	Prevalence of depression is 63.4%
15	Hein et al. (2007)	229 Palestinian adolescents/Gaza Strip Age: 15–19 Beck's Depression Inventory-II	Palestine (Gaza)	Prevalence rate of depression 40.0%
16	Kolltveit et al. (2012)	136 school adolescents Age:12–17 Depression Self-Rating Scale for Children (DSRS)	Palestine (Gaza)	Prevalence rate of depression 41.2%
17	Elbedour et al. (2007)	299 adolescent students Age: 15–19 Beck's Depression Inventory-II	Palestine (Gaza)	Prevalence rate of depression 40%
18	Shaaban and Baashar (2003)	1107 girls Age:12–19 Beck's Depression Inventory (BDI)	Sudan	The prevalence rate of depressive disorder 11%
19	Abdullatif (1995)	1981 adolescent students Age:12–18 Arabic Children's Depression Inventory	Kuwait	The prevalence rate of depression of boys 3.7% and girls 4.8%
20	Al-Balhan (2006)	2299 adolescents Age: 9–17 Children's Depression Inventory (CDI)	Kuwait	The mean depression score was below the cutoff for depression

(continued)

Table 1 (continued)

	Study	Data	Location	Findings
21	Modabernia et al. (2007)	3681 adolescents Age:14–20 Beck's Depression Inventory-II	Iran	34% suffered from depression
22	Moeini et al. (2019)	670 secondary school female students Age: 15–18 Persian version of Center for Epidemiologic Studies Depression Scale (CES-D)	Iran	The prevalence rate of depression 52.2%
23	Horesh et al. (2003)	65 adolescents Beck's Depression Inventory (BDI)	Israel	The prevalence rate of depression 32%

depression among adolescents were between 7.1% and 19.4% (Balazs et al. 2012), whereas the USA had prevalence rates of between 11.0% and 7.5% (Avenevoli et al. 2015).

Expectedly, around two third of participants from areas in which occupation and armed conflict were active, such as Palestine (Gaza), Iraq, and Israel, reported some form of depressive symptoms (Freh 2016; Horesh et al. 2003; Hein et al. 2007). This indicated that war substantially contributed to the mental health difficulties within the MENA region.

Overall, evidence has shown that depression is highly prevalent among adolescents from the MENA region. Although the prevalence of depression varied tremendously across studies, this region remained culturally and ethnically homogenous (Nydell 2005). Generally, individuals from the MENA region share almost similar outlooks toward mental health issues, which stem from their shared cultural and social norms (Nydell 2005). However, some variability in the results had been observed, perhaps due to case definition, study design, source of subjects, time frame (Desouky et al. 2015), and measurement methods. Given that depression is clinically heterogenous, understanding its pathogenesis among adolescents remains challenging considering the overlapping risk factors that greatly impact its occurrence (Thapar et al. 2012).

4.2 *Risk Factors for Depression*

Several risk actors have been known to promote the occurrence of depressive symptoms among adolescents from the MENA region, including gender, age, psychosocial contexts, family history, and political conflicts.

Gender

Most studies across the region have reported that females are 1.5–2 times more likely to develop depression than males (Kolltveit et al. 2012; Desouky et al. 2015; Al-Balhan 2006; Al Gelban 2009; Abdullatif 1995; Asal and Abdel-Fattah 2007; Afifi 2006; Dardas 2017; Modabernia et al. 2007; AlBuhairan et al. 2015). One possible explanation could be that females and males within the MENA region encounter enormous pressure to conform with gender roles imposed by the Arabic society. However, females encounter an even firmer pressure to control their behavior. For instance, a wide range of society-based sources help to define gender roles, including parents, family's norm, culture, peers, teachers, and the media (Priess et al. 2009). In the face of insurmountable pressure, female adolescents become inclined to fulfill their gender role identities, which presumably prepares them for their eventual adult role as a woman (Priess et al. 2009). On the other hand, males exercise a higher level of freedom, with the culture being more forgiving of their improprieties. Unlike female misconducts, those committed by males rarely taint the family's honor. Moreover, males are allowed to have and express stronger ideological and political commitments, which was found to be a protective factor in stressful reactions to political conflicts (Kolltveit et al. 2012).

Overall, disparities in the incidence of depression between females and males have been well documented from clinical samples and have been robust across different methods of assessment. Hence, gender differences in the rates of depression are unlikely to be caused by methodological differences or sample characteristics. As such, differences could be linked to female hormonal changes, which suggests that depression is directly linked to pubertal hormonal changes (e.g., testosterone) among females (Copeland et al. 2019). Animal studies have indicated that estrogen increases the stress response in the prefrontal cortex. Thus, post-pubertal sex differences in depression might be partly caused by increased exposure to stressors and hormonally linked heightened stress sensitivity among females (Thapar et al. 2012).

Age

Studies conducted in the MENA region have reported controversial data with regard to age. Accordingly, some studies have claimed that depressive symptoms increase around late childhood and early adolescence and typically persist into adulthood (Dardas 2017; Ismayilova et al. 2013; Obermeyer et al. 2015). A potential explanation could be the ability of contextual factors, such as gender, to mediate physical and psychological changes that often emerge around puberty and tend to greatly increase toward adulthood (Cook et al. 2009). For instance, throughout adolescence, females developed more self-consciousness and reported lower self-esteem compared to their male counterparts. Moreover, females were found to be more accommodating in their interpersonal relationship (Priess et al. 2009). Intersections between these contextual factors have been found to regulate social and cultural norms and hence influence the emergence of depressive symptoms according to age.

In addition, hormonal and maturational changes can alter an individual's mental health status (Thapar et al. 2012).

Psychosocial Context

The psychosocial profile, which includes parents' education, region of residence, and verbal and physical bullying (Dardas 2017; Asal and Abdel-Fattah 2007; Moeini et al. 2019; Afifi 2006; Afifi 2007; Ali et al. 2014), has been identified as a strong risk factor for depression among adolescents from the MENA region. Combinations of the aforementioned factors directly or indirectly expose adolescents to official and unofficial violence, thereby promoting an increasing sense of helplessness and despair (Hein et al. 2007; Abdel-Khalek 2009; Asal and Abdel-Fattah 2007), poor relationships with parents and colleagues, and lack of family support and family discordance (Al-Kaabi et al. 2017; Ali et al. 2014; Horesh et al. 2003), which have also been linked to higher levels of depressive symptoms. Interestingly, family dynamics have been identified as a crucial influencing factor on the physical and mental health of its members, with any impairments in such a dynamic impacting the family as a whole. Hence, levels of parent–adolescent conflict leads to a wide range of psychological difficulties (Lee et al. 2010). Furthermore, the presence of such disagreement indicates that the basic needs of adolescents have not been satisfied, which subsequently leads to anxiety, tension, and depression (Ferro and Boyle 2015). Additionally, low socioeconomic status, poverty, and economic difficulties within the MENA region have been considered antecedents for depression (Ismayilova et al. 2013; Ali et al. 2014; Modabernia et al. 2007; Moeini et al. 2019; Freh 2016; Hein et al. 2007; Kolltveit et al. 2012; Elbedour et al. 2007).

However, studies in Saudi Arabia have reported that socioeconomic status was unrelated to depression (Al-Gelban 2007). Instead, these studies suggested that increased stress levels among Saudi families were related to social and cultural transformations. Moreover, choosing a suitable career has become increasingly difficult as the job market for women has become more complex. Furthermore, high psychological distress among Saudi students has been related to academic problems during their secondary school period (AlBuhairan et al. 2015). This could be associated with the lifestyle transformations occurring in Saudi Arabia and UAE, the adaptation to which has accelerated their pace of life.

Furthermore, maltreatment from teachers, school bullying, classroom environment, increasing independence (Al-Gelban 2007), and discordant body image (Al-Amer et al. 2019) have been found to mirror the levels of depressive symptoms among adolescents. Hence, locus of control (i.e., individuals' beliefs regarding their ability to control daily reinforcements and cope with challenges) had been found to be a protective factor against depression among adolescents. Good locus of control allows adolescents to avoid bullying, which indicates some immunity against depression. One study showed that a history of abuse during adolescence predicted the occurrence of depression (Afifi 2006). Hence, the lack of internal control is highly likely to promote mental health problems, such as depressive symptoms

(Afifi 2007). Unlike external locus of control, which is determined by external power, internal locus of control allows adolescents to feel confident and reduce their levels of anxiety, which would consequently promote good mental health and decrease vulnerability toward depressive symptoms by removing anxiety symptoms considered to be antecedents for depression. Moreover, internal locus of control has been considered one of the common protective factors for child abuse (Afifi 2007).

Family History

A family history of chronic diseases and psychological illnesses had been identified as an important risk factor for depressive symptoms among adolescents (Asal and Abdel-Fattah 2007; Abou Abbas and AlBuhairan 2017; Dardas 2017). Studies have shown that physical health problems among family members reflect dysfunctional mental health status, such as sadness or hopelessness, which are amplified among females (Abou Abbas and AlBuhairan 2017). The family system theory defines a family as an organized unit in which all elements are interdependent such that each family member influences and is influenced by other members. Hence, it is plausible that adolescents in collectivist communities would be influenced by the physical and mental problems of another family member (Restifo and Bögels 2009). Moreover, the biological theory suggests that mental illness has biological implications; hence, adolescents who live with parents suffering from depression are more vulnerable to depressive symptoms.

Political Climate

A substantial amount of studies have investigated the association between depression and environmental factors, such as exposure to acute stressful events (Asal and Abdel-Fattah 2007) (e.g., personal injury and bereavement). Thus, some children exposed to the traumatic events of war, torture, displacement, and orphanhood are highly likely to develop long-term psychiatric disorders, especially among those who have a family history of such disorders or are exposed to multiple traumas (Thapar et al. 2012).

Considerable evidence has shown a significant association between depression symptoms among adolescents and exposure to armed conflict, including those occurring in Iraq (Freh 2016) and Palestine (Hein et al. 2007; Kolltveit et al. 2012) and Israel (Horesh et al. 2003). A systematic review reported high rates of mental health problems among Palestinians and Iraqis (Dimitry 2012). Established data obtained through structured clinical interviews and standardized questionnaires have shown that adolescents exposed to war are vulnerable to a wide range of mental health problems, such as post-traumatic stress disorder, anxiety and depression, and externalizing behaviors (i.e., hostility) (Buka et al. 2001). Political instability and poverty have been the most common cited causes for worsening daily stressors, as well as social and economic challenges, which in turn add pressure to overall psychological burden among war-affected populations. For instance, daily stressors may influence the development and implementation of interventions that could potentially improve a range of modifiable factors in post-conflict settings. A review suggested that those with comorbid depression and anxiety had an earlier age of onset of depression and that anxiety preceded depression in the majority of cases. Kovacs and Devlin (1998) suggested that comorbid anxiety and depression may be a single disorder in some children, whereas both disorders may be distinct and may indicate a higher level of vulnerability in others (Kovacs and Devlin 1998). Although a few studies have been available regarding comorbidity across developmental stages, evidence suggests that anxiety and depression are comorbid across the patient's lifespan (Hein et al. 2007).

4.3 Consequences

Studies have shown that depression and anxiety coexist within the MENA region (Desouky et al. 2015; Al-Gelban 2007; Al-Gelban et al. 2009; Hein et al. 2007). These comorbidities stem from the fact that anxiety is considered an antecedent for depression, both of which often overlap. Another serious implication for depression is suicide. For instance, one review reported that around 24% of adolescents had seriously considered attempting suicide at least once (Dardas et al. 2018b). However, a community study of 2349 Jordanian adolescents aged 12 to 17 reported lower rates of suicidal thought among females (8%) and males (7%) (Dardas et al. 2018a). This is inconsistent with international trends in which adolescent females exhibit more suicidal ideations and suicidal attempts than males, who are more prone to engage in high-risk behaviors and are more likely to commit suicide (Hawton et al. 2002; Parker and Roy 2001). Importantly, suicidal thoughts remain dangerous regardless of their prevalence. In fact, studies have identified suicide as the second leading cause of death among adolescents worldwide (Centers for Disease Control and Prevention 2010).

Depression has also been associated with risky behaviors among Arab adolescents, which were reported to be higher among adolescent males aged 14–16 years who reported ever drinking alcohol (Ismayilova et al. 2013). Furthermore, adolescents with more depressive symptoms were more likely to initiate smoking, smoke more, continue smoking, and have problems quitting smoking (Afifi 2007). Moreover, they were more likely to have reported being involved in fighting and/or bullying behaviors (Celedonia et al. 2013). On the other hand, one study showed that having a hobby, eating breakfast regularly, not smoking, getting enough sleep at night, and engaging in physical exercises protected them against depressive symptoms (Afifi 2006). These findings were similar to those reported in Western literature (Cairns et al. 2014).

5 Summary and Recommendations

Depression is a chronic debilitating disease that is widely prevalent among adolescents from the MENA region. Studies have shown that depression during the early stages of life continues and often reoccurs in many cases during adulthood. Moreover, depression can be characterized as a clinically heterogenous disease wherein a wide range of risk factors contribute to its development, including female sex, age, regional political conflicts, low socioeconomic status, and poverty. Depression also leads to serious psychosocial impairments. Hence, early detection and treatment of depression are imperative. Mental health problems, particularly depression, have still not been a prime focus for policymakers, researchers, and health providers (Afifi 2007). Hence, establishing mental health services and effective interventions for adolescents remains important. Furthermore, prevention strategies may need to be initiated at a younger age to account for the rise in depression symptoms among adolescence (DARDAS 2017). Preventing the onset and recurrence of depression in this vulnerable population by recognizing and treating this disorder cannot be understated. Preventive methods among those with high familial risk should include effective treatment of parental depression and irritability and fear/anxiety among children, as well as strategies for addressing social risk factors. Family-based programs may be indicated for children at high familial risk of depression given that parental depression is associated with social adversity (poverty and stress exposure) and moderates the effectiveness of preventive programs focusing on children.

Considering the inconsistencies among studies involving the MENA region and the many negative outcomes of depression, estimating the prevalence of depressive symptoms is imperative to design effective preventive strategies aimed at this vulnerable age group. Given that most of the self-administered questionnaires cannot directly diagnose depression, the prevalence of depression should be investigated using standard interviews based on the Structured Clinical Interview for DSM Disorders (5) and the Composite International Diagnostic Interview, which can establish a clinical diagnosis according to the Statistical Manual of Psychiatric Disorders, 5th edition (DSM-5).

Critical Thinking/Reflection Questions

1. Reflect on why adolescents in the MENA region are considered more vulnerable to depression in comparison to their counterparts from the developed world.
2. Discuss the possible reasons for the lack of mental health counseling services among collectivist communities.
3. Reflect on the discrepancies in the prevalence of depression among adolescents in the MENA region although they are culturally and ethnically a homogenous community.
4. How the highly interdependent nature of families in the MENA region influences the levels of depressive symptoms among adolescents?
5. Identify and analyze the protective factors against depression among adolescents in the MENA region.

Acknowledgments I would like to thank Dr. Motasem Darwish for his help in the data extraction in this chapter.

References

Abdel-Khalek, A. M. 2009. Religiosity, subjective well-being, and depression in Saudi children and adolescents. *Mental Health, Religion & Culture,* 12, 803-815.

Abdullatif, H. I. 1995. Prevalence of depression among middle-school Kuwaiti students following the Iraqi invasion. *Psychological reports,* 77, 643-649.

Abou Abbas, O. & Albuhairan, F. 2017. Predictors of adolescents' mental health problems in Saudi Arabia: findings from the Jeeluna® national study. *Child and adolescent psychiatry and mental health,* 11, 52.

Afifi, M. 2006. Depression in adolescents: gender differences in Oman and Egypt. *EMHJ-Eastern Mediterranean Health Journal, 12 (1-2), 61-71, 2006.*

Afifi, M. 2007. Health locus of control and depressive symptoms among adolescents in Alexandria, Egypt.

Al-Balhan, E. M. 2006. The Children's Depression Inventory as a reliable measure for post-Iraqi invasion Kuwaiti youth. *Social Behavior and Personality: an international journal,* 34, 351-366.

Al-Gelban, K. S. 2007. Depression, anxiety and stress among Saudi adolescent school boys. *The journal of the Royal Society for the Promotion of Health,* 127, 33-37.

Al-Gelban, K. S., Al-Amri, H. S. & Mostafa, O. A. 2009. Prevalence of depression, anxiety and stress as measured by the depression, anxiety, and stress scale (DASS-42) among secondary school girls in Abha, Saudi Arabia. *Sultan Qaboos University Medical Journal,* 9, 140.

Al-Kaabi, N., Selim, N., Singh, R., Almadahki, H. & Salem, M. 2017. Prevalence and Determinants of Depression among Qatari Adolescents in Secondary Schools. *Family Medicine & Medical Science Research,* 6.

Al-Amer, R., Salamonson, Y., Villarosa, A. R., Subih, M., Darwish, R. & Maneze, D. 2019. Accuracy of Body Weight Estimation Among Palestinian Refugee Adolescents Living in Jordan: A Cross-Sectional Study. *Journal of Nursing Scholarship,* 51, 642-650.

Al Gelban, K. S. 2009. Prevalence of psychological symptoms in Saudi secondary school girls in Abha, Saudi Arabia. *Annals of Saudi medicine,* 29, 275-279.

Albuhairan, F. S., Tamim, H., Al Dubayee, M., Aldhukair, S., Al Shehri, S., Tamimi, W., El Bcheraoui, C., Magzoub, M. E., De Vries, N. & Al Alwan, I. 2015. Time for an adolescent health surveillance system in Saudi Arabia: findings from "Jeeluna". *Journal of Adolescent Health,* 57, 263-269.

Ali, S., Al Faisal, W., Mahdy, N. H., Hussein, H., El Sawaf, E. & Wasfy, A. 2014. Determinants and prevalence of Depressive Symptoms in Secondary School Students in Dubai. *Middle East Journal of Psychiatry & Alzheimers,* 5.

Asal, A. & Abdel-Fattah, M. M. 2007. Prevalence, symptomatology, and risk factors for depression among high school students in Saudi Arabia. *Neurosciences,* 12, 8-16.

Avenevoli, S., Swendsen, J., He, J.-P., Burstein, M. & Merikangas, K. R. 2015. Major depression in the national comorbidity survey–adolescent supplement: Prevalence, correlates, and treatment. *Journal of the American Academy of Child & Adolescent Psychiatry,* 54, 37-44. e2.

Balazs, J., Miklósi, M., Keresztény, Á., Apter, A., Bobes, J., Brunner, R., Corcoran, P., Cosman, D., Haring, C. & Kahn, J.-P. 2012. P-259-Prevalence of adolescent depression in Europe. *European Psychiatry,* 27, 1.

Bener, A., Dafeeah, E. E., Chaturvedi, S. K. & Bhugra, D. 2013. Somatic symptoms in primary care and psychological comorbidities in Qatar: neglected burden of disease. *International Review of Psychiatry,* 25, 100-106.

Buka, S. L., Stichick, T. L., Birdthistle, I. & Earls, F. J. 2001. Youth exposure to violence: Prevalence, risks, and consequences. *American journal of orthopsychiatry,* 71, 298-310.

Cairns, K. E., Yap, M. B. H., Pilkington, P. D. & Jorm, A. F. 2014. Risk and protective factors for depression that adolescents can modify: a systematic review and meta-analysis of longitudinal studies. *Journal of affective disorders,* 169, 61-75.

Celedonia, K. L., Wilson, M. L., El Gammal, H. A. & Hagras, A. M. 2013. Physical fighting among Egyptian adolescents: social and demographic correlates among a nationally representative sample. *PeerJ,* 1, e125.

Centers for Disease Control and Prevention. 2010. *Injury Prevention & Control: WISQARS™ — Web-based Injury Statistics Query and Reporting System* [Online]. Available: https://www.cdc.gov/injury/wisqars/index.html [Accessed October,30 2020].

Cook, M. N., Peterson, J. & Sheldon, C. 2009. Adolescent depression: an update and guide to clinical decision making. *Psychiatry (Edgmont),* 6, 17.

Copeland, W. E., Worthman, C., Shanahan, l., Costello, E. J. & Angold, A. 2019. Early pubertal timing and testosterone associated with higher levels of adolescent depression in girls. *Journal of the American Academy of Child & Adolescent Psychiatry,* 58, 1197-1206.

Csikszentmihalyi, M. 2020. *Adolescence* [Online]. Encyclopædia Britannica. Available: https://www.britannica.com/science/adolescence [Accessed October 20, 2020].

Dardas, L. A. 2017. *A nationally representative survey of depression symptoms among Jordanian adolescents: Associations with depression stigma, depression etiological beliefs, and likelihood to seek help for depression.* Duke University.

Dardas, L. A., Silva, S., Noonan, D. & Simmons, L. A. 2018a. Studying depression among Arab adolescents: Methodological considerations, challenges, and lessons learned from Jordan. *Stigma and Health,* 3, 296.

Dardas, L. A., Silva, S. G., Smoski, M. J., Noonan, D. & Simmons, L. A. 2017. Adolescent depression in Jordan: Symptoms profile, gender differences, and the role of social context. *Journal of psychosocial nursing and mental health services,* 56, 44-55.

Dardas, L. A., Silva, S. G., Smoski, M. J., Noonan, D. & Simmons, L. A. 2018b. The prevalence of depressive symptoms among Arab adolescents: findings from Jordan. *Public Health Nursing,* 35, 100-108.

Desouky, D., Ibrahem, R. A. & Omar, M. S. 2015. Prevalence and comorbidity of depression, anxiety and obsessive compulsive disorders among Saudi secondary school girls, Taif Area, KSA.

Dimitry, L. 2012. A systematic review on the mental health of children and adolescents in areas of armed conflict in the Middle East. *Child: care, health and development,* 38, 153-161.

El-Islam, M. F. 1982. Arabic cultural psychiatry. *Transcultural psychiatric research review,* 19, 5-24.

Elbedour, S., Onwuegbuzie, A. J., Ghannam, J., Whitcome, J. A. & Hein, F. A. 2007. Post-traumatic stress disorder, depression, and anxiety among Gaza Strip adolescents in the wake of the second Uprising (Intifada). *Child Abuse & Neglect,* 31, 719-729.

Ferro, M. A. & Boyle, M. H. 2015. The impact of chronic physical illness, maternal depressive symptoms, family functioning, and self-esteem on symptoms of anxiety and depression in children. *Journal of abnormal child psychology,* 43, 177-187.

Freh, F. M. 2016. PTSD, depression, and anxiety among young people in Iraq one decade after the American invasion. *Traumatology,* 22, 56.

Ghareeb, G. A. & Beshai, J. 1989. Arabic version of the Children's Depression Inventory: reliability and validity. *Journal of Clinical Child Psychology,* 18, 323-326.

Hawton, K., Rodham, K., Evans, E. & Weatherall, R. 2002. Deliberate self harm in adolescents: self report survey in schools in England. *Bmj,* 325, 1207-1211.

Heckman, A. R. 2018. Jewish Radicals of Morocco: Case Study for a New Historiography. *Jewish Social Studies,* 23, 67-100.

Hein, F. A., Elbe, S., Onwuegbuzie, A., Ghannam, J. & Whitcome, J. 2007. Posttraumatic stress disorder, depression, and anxiety among Gaza strip adolescents in the wake of the second uprising (Intifada). *Al Aqsa University,* 11.

Horesh, N., Orbach, I., Gothelf, D., Efrati, M. & Apter, A. 2003. Comparison of the suicidal behavior of adolescent inpatients with borderline personality disorder and major depression. *The Journal of nervous and mental disease,* 191, 582-588.

Ismayilova, L., Hmoud, O., Alkhasawneh, E., Shaw, S. & El-Bassel, N. 2013. Depressive symptoms among Jordanian youth: results of a national survey. *Community mental health journal,* 49, 133-140.

Khouri, R. G. 2019. Poverty, inequality and the structural threat to the Arab region. *Shifting Global Politics and the Middle East*, 28.

Kolltveit, S., Lange-Nielsen, I. I., Thabet, A. A. M., Dyregrov, A., Pallesen, S., Johnsen, T. B. & Laberg, J. C. 2012. Risk factors for PTSD, anxiety, and depression among adolescents in Gaza. *Journal of traumatic stress,* 25, 164-170.

Kovacs, M. & Devlin, B. 1998. Internalizing disorders in childhood. *Journal of Child Psychology and psychiatry,* 39, 47-63.

Lee, A., Hankin, B. L. & Mermelstein, R. J. 2010. Perceived social competence, negative social interactions, and negative cognitive style predict depressive symptoms during adolescence. *Journal of clinical child & adolescent psychology,* 39, 603-615.

Lewinsohn, P. M., Rohde, P., Seeley, J. R., Klein, D. N. & Gotlib, I. H. 2000. Natural course of adolescent major depressive disorder in a community sample: predictors of recurrence in young adults. *American Journal of Psychiatry,* 157, 1584-1591.

Liu, Q., He, H., Yang, J., Feng, X., Zhao, F. & Lyu, J. 2020. Changes in the global burden of depression from 1990 to 2017: Findings from the Global Burden of Disease study. *Journal of psychiatric research,* 126, 134-140.

Lopez, A. D., Mathers, C. D., Ezzati, M., Jamison, D. T. & Murray, C. J. 2006. *Global burden of disease and risk factors*, The World Bank.

Modabernia, M., Shodial, T. H., Moosavi, S., Jahanbakhah, A. & Falahi, M. 2007. The prevalence of depression among high school and preuniversity adolescents: Rasht, northern Iran.

Moeini, B., Bashirian, S., Soltanian, A. R., Ghaleiha, A. & Taheri, M. 2019. Prevalence of depression and its associated sociodemographic factors among Iranian female adolescents in secondary schools. *BMC psychology,* 7, 25.

Nydell, K. 2005. Understanding Arabs: A Guide for Modern Times. Nicholas Brealey Publishing, London, Print.

Obermeyer, C. M., Bott, S. & Sassine, A. J. 2015. Arab adolescents: health, gender, and social context. *Journal of Adolescent Health,* 57, 252-262.

Parker, G. & Roy, K. 2001. Adolescent depression: a review. *Australian & New Zealand Journal of Psychiatry,* 35, 572-580.

Priess, H. A., Lindberg, S. M. & Hyde, J. S. 2009. Adolescent gender-role identity and mental health: Gender intensification revisited. *Child development,* 80, 1531-1544.

Restifo, K. & Bögels, S. 2009. Family processes in the development of youth depression: Translating the evidence to treatment. *Clinical Psychology Review,* 29, 294-316.

Shaaban, K. & Baashar, T. 2003. A community study of depression in adolescent girls: prevalence and its relation to age. *Medical principles and practice,* 12, 256-259.

Thapar, A., Collishaw, S., Pine, D. S. & Thapar, A. K. 2012. Depression in adolescence. *The Lancet,* 379, 1056-1067.

UNICEF 2011. *The state of the world's children 2011-executive summary: Adolescence an age of opportunity.*, Unicef.

World Health Organization. 2005. *Atlas : child and adolescent mental health resources : global concerns: implications for the future* [Online]. Available: https://apps.who.int/iris/handle/10665/43307 [Accessed 30 June 2020].

‘Cn I jus txt, coz I don wan 2b heard’: Understanding Mental Illness Stigma in Arab Youth’s Everyday Lives

Patricia Niland

1 Introduction

Anxiety and depression are mental health challenges that can profoundly impact young people’s lives, leading to social withdrawal, isolation, loneliness and suicide (WHO 2020a). Globally, depression is the fourth leading cause of adolescent illness and disability, and anxiety is the sixth leading cause for 10–14-year-olds (WHO 2020a). Mental illness stigma – the fear, rejection, avoidance and discrimination against people with mental illness – exists globally among all cultures and socio-demographic groups (Larrier et al. 2017). It is one of the toughest barriers for people to seek help, often creating more suffering and isolation than the mental problem itself. Mental illness stigma is recognised globally as a barrier for young people to seek help for mental health problems (WHO 2020b) making them vulnerable to discrimination, social exclusion, educational difficulties, physical ill-health and human rights violations (WHO 2020b). Arab youth experience mental health conditions such as anxiety and depression in their everyday lives but silence these issues due to societal and cultural stigma (Maalouf et al. 2019; Zolezzi et al. 2018). Arab youth in 15 MENA countries have now broken this silence, stating that mental illness stigma is pervasive in their everyday lives (Arab Youth Survey 2019). Ways to investigate and help address mental health conditions in Arab youth, such as anxiety and depression, and mental illness stigma are needed.

P. Niland (✉)
Accident Compensation Corporation, Auckland, New Zealand
e-mail: Patricia.niland@acc.co.nz

C. Barakat, F. Al Anouti (eds.), *Adolescent Mental Health in The Middle East and North Africa*, Global Perspectives on Health Geography,
https://doi.org/10.1007/978-3-030-91790-6_4

In this chapter you will firstly gain an understanding of the concept of mental illness stigma as a social determinant of mental health. This public health approach shifts the focus from 'individual patients' being to blame for their mental issues, to considering environmental factors including interactions between societal (family, social networks, community and neighbourhood), economic, political and cultural conditions overall. Secondly, a rapid literature review of research exploring mental illness stigma for Arab youth is reported, showing the barriers this stigma sets up for them to seek help. Thirdly, the challenges faced by Arab mental health researchers due to mental illness stigma are explored and suggestions are offered for ways to engage with Arab young people to combat this stigma.

2 Section 1: What Is Mental Illness Stigma

2.1 What Is Mental Health?

According to the World Health Organization (WHO), mental health is 'a state of well-being in which the individual realizes his or her own abilities, can cope with the normal stresses of life, can work productively and fruitfully, and is able to make a contribution to his or her community' (WHO 2020c, p.1). This definition is consistent across cultures and reflects a focus on positive mental health as a foundation for well-being for a person within their everyday worlds.

It is recognised, however, that people's mental health is shaped by multiple and interacting biological, psychological, environmental and social factors (Engel 1977; biopsychosocial model). A way to think about this is the Health Onion model (Dahlgren and Whitehead 1991). This model is holistic – it takes account of a person's health as having many interconnected layers (like an onion). At the core an individual's health is influenced by community, living and working conditions and the socioeconomic, political, cultural and physical environment. In this model, a person's age, gender and hereditary are individual factors, and their lifestyle choices (e.g. healthy diet, exercise) impact on their health. Level 2 is social and community influences (e.g. family and friends). Level 3 is a person's living and working conditions (e.g. housing, employment, education); and level 4 is the general socioeconomic, cultural, political and environmental conditions (e.g. religion, government and healthcare systems). Each of these spheres of influence interacts to shape or determine a person's health.

These influences or conditions are social determinants of health because where there are restrictions or negative influences in these spheres (e.g. poverty), a person's health can be impeded. Positive influences and conditions (e.g. free medical care) in these spheres can promote good health outcomes. Mental illness stigma is a social determinant of health because it is a socio-cultural influence that may impede well-being, including positive mental health outcomes for people.

2.2 *Mental Illness Stigma Is a Social Determinant of Mental Health*

The World Health Organization (WHO 2020c) defines the social determinants of health as:

> ...the conditions in which people are born, grow, work, live, and age, and the wider set of forces and systems shaping the conditions of daily life. These forces and systems include economic policies and systems, development agendas, social norms, social policies and political systems.

The *conditions* in which people are born, grow, work, live and age include early life experiences, education, employment, income, housing, neighbourhood and built environment, food and nutrition and access to healthcare. When these living conditions are optimal, they determine positive physical and mental health development throughout our lives. Marmot and Wilkinson (2006) have summarised the global evidence for the role of lack of access to, or negative experiences of, these living conditions in the development of disease and the persistence of health inequalities. Lack of access to positive living conditions (e.g. an unhealthy diet or unsafe housing) is therefore identified as a risk factor for negative physical and mental health outcomes. Risk factors need to be understood within the broader societal context that shapes them.

As the WHO definition states, conditions of daily life are shaped by *political systems, social policies and social norms*. These societal systems shape living conditions by determining the priorities and allocations of resources, such as education and income, to the population. When people cannot access these resources, or they are below standard, they experience negative living conditions that in turn become risk factors for poor physical and mental health outcomes. To reduce negative mental health outcomes, in particular, requires us to look beyond the individual to the political, social, economic and cultural environments that shape the living conditions that determine our well-being. The public health approach to mental health is grounded in this conceptualisation (way of thinking about) of mental health being largely determined by the environment (i.e. societal living conditions). With this 'big picture' in mind we can explore what it means when we say mental health stigma is a social determinant of mental health. We will firstly consider the psychology of the *stigma process* and then look more specifically at mental illness stigma.

2.3 *The Stigma Process*

Sewilam et al. (2015) portray stigma as a combination of attitudes, stereotypes, prejudice and discrimination. **Attitudes** are our beliefs, values and emotional reactions to the world around us, formed through experiences and predisposing us to react to ourselves, other people, places, things and events in certain ways. **Stereotypes** are attitudes about groups of people based on general beliefs that are often inaccurate. **Prejudice** is a negative attitude towards particular groups based on derogatory stereotypes. **Discrimination** occurs when people act on the basis of prejudiced attitudes. For instance, a person may believe that all criminals are dangerous (a stereotype) and distrust them because of their presumed lack of control (prejudice) so they maintain a social distance from them or will not employ them, or accept them into their families (discrimination). The stigma process describes the transformation of attitudes to discrimination via creation of stereotypes that lead to prejudice. The next section sets out how this stigma process in relation to mental illness is enacted through **public stigma** and **self-stigma** practices.

2.4 Mental Illness Public Stigma and Self-Stigma

Public mental illness stigma occurs when individuals endorse stereotypes about mental illness and act on their beliefs in these stereotypes to discriminate against other individuals (Sewilam et al. 2015). The ensuing prejudice (biased negative attitudes) leads to discrimination. Discrimination occurs when an individual or group blocks people's access to conditions of daily living (e.g. employment, housing and marriage) because of a negative attitude, for example, when a person who is considered mentally ill is not considered for a job.

Self-stigma occurs when an individual internalises others' negative ideas and responses, leading to negative thoughts and emotional reactions towards themselves (Corrigan and Watson 2002; Dardas et al. 2017; Goffman 1963). Mental illness self-stigma is experienced as fear and shame so that people silence their mental health challenges (Dardas et al. 2019a; Zolezzi et al. 2018). A significant harm of self-stigma is the shutting down of life goals such as education and employment due to low self-esteem and self-efficacy ('Why try?') because society deems mentally ill people cannot succeed in these important life pursuits (Corrigan et al. 2009). Another significant harm is that people with mental illness self-stigma avoid seeking help for mental health problems to avoid negative labels, knowing the 'mental illness' label will engender public disrespect and devalue them (Corrigan and Rao 2012).

The next section sets out the literature specific to mental health stigma for young people in the MENA region. A rapid review (Garritty et al. 2020) of the published peer-reviewed qualitative literature was undertaken, identifying three systematic reviews (Dardas et al. 2017; Gearing et al. 2013; Zolezzi et al. 2018) and seven narrative reviews (Al-Krenawi 2005; Corrigan and Rao 2012; Corrigan et al. 2009; Corrigan and Watson 2002; Maalouf et al. 2019; Merhej 2019; Sewilam et al. 2015) that met robust quality criteria (Hannes 2011). Search terms included: adolescent, young people, mental health, stigma, barrier, Arab, MENA, depression and anxiety. No restrictions were placed on the time period. Searched databases included Web of Science, Scopus, OVID, EBSCO, Google Scholar and Cochrane. Studies that focussed on adults, severe psychiatric disorders, non-Arab peoples and non-stigma mental health issues were excluded.

When reading this section, keep in mind the concepts of public stigma and self-stigma, and the Health Onion model that places a person within a holistic context of social life, and the broader socioeconomic, cultural, environmental and political influences that people draw on to make sense of mental conditions such as anxiety and depression. The research identified draws out the experiences of mental illness stigma for young adults in the MENA region; however, it is clear that this research is limited by mental illness public and self-stigma which sets up barriers (e.g. funding, identification of participants) to this investigation. An alternative approach is needed that engages with Arab young people within their everyday worlds.

3 Section 2: MENA Mental Illness Stigma Rapid Review

Mental health conditions exist for Arab youth and there is recognition of public and self- stigma for experiencing anxiety and depression. The Arab Youth Survey (2019) included 'mental health' questions for the first time in its 11 years of publication. This suggests that social and cultural attitudes to mental health are beginning to be talked about openly in the Arab world. The 3300 face-to-face interviews with Arab youth in 15 MENA countries for the Arab Youth Survey identified that 31% knew of someone with mental health issues, such as anxiety or depression, but 69% reported they did not know of anyone. Youth respondents were divided over whether a mental illness stigma exists; 49% viewed mental illness as a normal part of life but 50% viewed it negatively. These findings call for further investigation into how Arab youth view and make sense of their mental health.

3.1 Anxiety and Depression

Arab youth report they have mental health problems such as anxiety and depression, and they are triggered by stress in personal relationships, lack of national security and safety, academic concerns, financial concerns, family issues and job-related stress (Arab Youth Survey 2019). A longitudinal survey of over 15,000 Egyptian young people (aged 10–29 years) found increased psychiatric disturbance, nervousness, uncertainty and reduced self-worth due to lower school achievement, being out of work, failure to marry and find a job and failure to finish school (Liu et al. 2017). Stresses in young Arab adults around everyday life situations relate to level 2 (social and community influences, e.g. family, community and friends) and level 3 (a person's living and working conditions, e.g. housing, employment, education) of the Health Onion model. As noted earlier, these layers are integral to a person's health; they are ingredients of a person's holistic health experience. The anxiety and depression that may result due to this stress are largely silenced due to the stigma associated with mental health issues (e.g. Zolezzi et al. 2018). In this sense, mental illness stigma is a negative social determinant because it is a barrier to achieving positive health outcomes and life experiences. This stigma is not only hidden, it is also deeply embedded in the socio-cultural beliefs and identity of Arab youth which influences their sense of day-to-day self.

3.2 Mental Illness and Negative Beliefs

For those who hold a stigmatised (negative) view of mental illness, it often derives from socio-cultural beliefs that mental illness is caused by personal weakness, spiritual practices, dangerous behaviours, drug use, or genetic factors, a condition that will negatively impact marriage and job prospects (Dardas et al. 2016). In their mental illness stigma literature review of 10 years of research in 16 MENA region countries (including mental health patients, care providers, family members, students), Zolezzi et al. (2018) found that although influenced by factors like education, gender and familiarity with mental illness, negative beliefs about the origins of mental illness were heavily influenced by religious teachings, cultural stereotypes and family traditions. Stigmatising beliefs included: 'God's punishment, God's will, evil eye, demons, spirits, paranormal phenomena, supernatural power, curse and magic…' (Zolezzi et al. 2018, p.600). These beliefs derive from the outer layer of the Health Onion model – the broader societal and cultural belief systems that people draw on to make sense of their identities. These socio-cultural beliefs create a public stigma for mental illness that, in turn, induces self-stigma feelings of low self-esteem, self-efficacy, and fear and shame for experiencing mental illness.

For instance, Zolezzi et al. (2018) found that mental illness was reported by Arab youth and healthcare professionals as being shameful, an embarrassment, and incompatible with true friendships, marriage or having children. Thus, the impact of self-stigma induces pervasive feelings of fear and shame and beliefs that being labelled as mentally ill will limit everyday practices such as marriage and job prospects (Dardas et al. 2016). On a basic behavioural level, Zolezzi et al. (2018) found that social distancing (unwilling to work closely with or be a neighbour) from people with mental illness was the most prevalent stigmatising action reported.

3.3 Mental Illness Stigma and Help-Seeking

In terms of help-seeking, a qualitative analysis of how Jordanian adolescents make sense of depression found they fear being labelled as mentally ill. This fear of a mentally ill label contributes to poor engagement in, adherence to and use of mental health services (Dardas et al. 2019a) as well as reluctance to seek professional help (Maalouf et al. 2019). Maalouf et al. (2019), in their action plan to reduce mental health stigma in the Arab region, recommend considering Arab people's willingness to seek informal help (e.g. a religious consultant, close friend or family member) rather than formal help because of the public stigma of seeing a psychiatrist.

The social or public shame of mental illness stigma may explain why people may firstly seek help within their families. Dardas et al. (2019a) investigated whether depression stigma impacted Jordanian adolescents' willingness to seek help and the type of treatment they would seek. In collaboration with the Jordanian Ministry of Education, they conducted a nationally representative, school-based survey of adolescents aged 12–17 years ($N = 2349$). One fourth of the adolescents with higher depression scores (>29, BDI-II; Beck et al. 1996) reported they would not seek professional help for depression. For adolescents willing to seek help, the most likely sources included family member (57%), school counsellor (46%), psychiatrist (43%), religious leader (39%) and general health practitioner (28%). The authors concluded that 'given the importance of family in Arab culture, it was not surprising that family was the most commonly cited source of help' (p.121). However, the family is also impacted by mental health stigma. Zolezzi et al. (2018) found that in the majority of mental illness stigma literature, it was identified that Arabs prefer seeking faith healers or connection to God to overcome mental illness. Reasons for not seeking mental health services were linked to stigmatised beliefs that mental illness harmed the family's reputation and feeling shame for having mental illness in the family. Their review also found that admission of family members to psychiatric services produces a stigmatising label for all family members. Similarly, Gearing et al. (2013), in their systematic review investigating barriers to and strategies for mental health interventions in Middle Eastern Arab countries, found that help-seeking for mental illness had negative consequences of social shame for using mental health services – families risked a damaged reputation or diminished social status in their communities.

This impact on individuals and families suggests that mental health education needs to focus on public awareness of 'the nature, natural course and treatments available for mental illnesses, targeting misleading associations of shame, blame, punishment and supernatural powers, but without undermining the positive messages of religious teachings that instil hope regarding treatment and prognosis' (Zolezzi et al. 2018, p.606).

Mental health education is best informed by evidence-based research that explores how Arab young people make sense of mental health. The next section discusses the challenges of doing mental health research in the MENA region. It is shown how mental illness stigma impedes this research and why it is important to explore Arab young people's own sense-making of mental health. A New Zealand youth mental health initiative is outlined to offer a way forward to engage with young people in their social media worlds.

4 Section 3: Mental Illness Stigma Research: A Way Forward

Education and interventions to address mental illness stigma are impeded by a lack of knowledge on how Arab youth make sense of well-being challenges such as anxiety and depression (Dardas et al. 2017). A lack of published intervention research in this space is due in part to the prevalence of mental illness stigma. Maalouf et al. (2019) report on an evidenced-informed call for action to identify challenges faced by Arab mental health researchers. This symposium included Arab mental health researchers, institutional and funding agency officials and international research collaborators. It was noted that: 'Although mental disorders are a leading cause of disability in the Arab region, which includes 5·54% of the global population, Arab countries produce only 1·0% of the global output of peer-reviewed publications in mental health research' (p.961). Research challenges identified by this group included 'recruitment barriers such as the stigma of seeking professional help and reluctance to self-disclose, to broader issues of scarce institutional resources, and insufficient mental health research expertise' (Maalouf et al. 2019). The authors conclude that major barriers to research recruitment are stigma, reluctance to self-disclose and low use of formal services. Similarly, professional, public and regional funding bodies are reluctant to engage with mental health research due to stigma and less urgency than other health conditions.

With these challenges in mind, research that investigates Arab young people's own meaning-making is fundamental to designing effective mental health interventions. An effective intervention must engage Arab youth in their everyday context. If we can learn more about the challenges and tensions they face, we can try to address the issue from their perspectives, taking account of their family and socio-cultural context. This cultural and religious and environmental perspective would also take account of Arab values and beliefs and indigenous health practices. Research that translates Arab youth perspectives into culturally appropriate interventions can provide guidance and resources for Arab mental health practitioners (Alean Al-Krenawi 2005). Further, beyond society and culture, mental illness stigma persists in Arab national primary healthcare through the absence of human rights-based mental health legislation and policies and lack of awareness of people's fear to seek mental healthcare (Merhej 2019). To investigate Arab young people's own meaning-making would inform mental health legislation and policies to engage meaningfully with them in intervention initiatives.

4.1 Making Sense of Depression

A recent study investigated Arab adolescents' experiences of depression and their attitudes towards depression interventions (Dardas et al. 2019b). The authors conducted a qualitative thematic (focus group) study to capture Jordanian adolescents' experiences of depression and attitudes towards interventions. Participants were 92 adolescents aged 14–17 years in 12 Jordanian schools who reported experiencing mild to moderate depressive symptoms. Two main themes (ways they made sense of depression) were identified: 'being a depressed adolescent' and 'living with depression'. To experience 'being a depressed adolescent' was made sense of as: 'feeling down', 'feeling frustrated', 'staying alone' and 'feelings of emptiness' (p.39). To 'live with depression' was made sense of as a fear of labelling: the fear of being called majnoon (crazy), being ashamed, misunderstood and discriminated against or estranged from their families if they voiced their emotions. They explained that this fear leads them to avoid seeking help from their parents (often perceived as the problem) and health experts. However, they were positive about using an anonymous Internet-based depression prevention and treatment program as long as their parents did not know they were doing so. It is not surprising that Arab young people are comfortable with online interventions. Nine out of ten young Arabs use at least one social media channel every day; half of them get their news on Facebook daily; YouTube is highly used (e.g. 77% of Egyptian millennials watch it every day); Snapchat is ubiquitous (e.g. 55% of 16–24-year-olds in Saudi Arabia use it daily); and 55% of MENA region millennials spend over 3 hours per day on video and they are twice as likely as their global counterparts to post content online (Radcliffe and Abuhmaid 2020). This approach to engaging with young people within their everyday social media worlds has been taken up by the New Zealand government's mental health services for young people.

4.2 A New Zealand Social Media Mental Health Intervention

In New Zealand, the Youth Mental Health Project was set up to combat New Zealand's high youth mild to moderate mental health problems (Gluckman and Hayne 2011; Social Wellbeing Agency 2018). The project identified that mental illness stigma could be usefully addressed by engaging with young people's online worlds (Malatest International 2016). The project developed evidence-based programmes to improve young people's resilience, identify mental health problems as early as possible and provide effective, youth-friendly and timely treatment. A number of programmes were initiated within communities, schools and primary care settings. These programmes included youth resilience sessions to teach strategies and skills to cope with difficult and challenging times in their lives; check and connect mentors (older students) to work with youth and their family to help them achieve their goals; and school guidance counsellors.

As part of an evaluation of the project, an online survey was completed by more than 3000 secondary students in NZ (Superu 2017). The results provided useful information on what can stress or support student well-being. They found that 94% had someone either inside or outside school to turn to if they were upset but only 68% would ask friends and/or family. They were too embarrassed or shy to go to school counsellors as other children would see them going to the office and bully and tease them about being crazy (majnoon). However, having access to online or telephone services was found to be particularly important for youth who would not ask anyone for help inside or outside of school.

4.3 *Youthline: 'Cn I jus txt, coz I don wan 2b heard'*

A NZ youth counselling service (Youthline 2017) provides free online and telephone counselling for young people aged 12–25 years old. Young people can text for advice and support when undergoing stress in all areas of their lives, not just traumatic events. Texting is anonymous, non-judgemental and listening-focussed, staffed by volunteer trained counsellors. A qualitative interview study of 21 young people who had used the Youthline texting service identified that they valued the privacy, autonomy and control they experienced using the service (Gibson and Cartwright 2014). The authors suggest that 'text counselling may help young people balance their contradictory needs for autonomy and connection and facilitate their engagement with counselling support' (p.96). An evaluation of the Youthline texting service identified that text volumes were high and increasing; texting conversations reached significant depth including relationship issues, mental health issues and safety concerns including intentions of suicide or self-harm; and anonymity was highly valued by young people as evidenced by this message: 'Cn I jus txt, coz I don wan 2b heard' (Gibson and Cartwright 2014, p.38). This approach to engaging with young people within their everyday social media worlds suggests a way forward to engage with Arab young people in their mental health challenges.

5 Summary

In this chapter you firstly gained an understanding of mental illness stigma as a social determinant of mental health. This public health approach shifts the focus from 'individual patients' being to blame for their mental problems, to considering environmental factors including interactions between societal (family, social networks, community and neighbourhood), economic, political and cultural conditions overall. Secondly, a rapid literature review of research exploring mental illness stigma for Arab youth was reported, showing the barriers this stigma sets up for them to seek help. Thirdly, the challenges faced by Arab mental health researchers due to mental illness stigma were explored and suggestions offered for ways to engage with Arab young people to combat this stigma. The New Zealand experience is an example of engaging with young people within their everyday online social worlds. This type of approach is a way to expand understanding of mental illness stigma and develop interventions to support Arab youth.

Critical Thinking/Reflection Questions

1. What is the Health Onion model of health?
2. What does it mean when we say that mental illness stigma is a social determinant of mental health?
3. How does public mental illness stigma differ from self-mental illness stigma?
4. What is your view of the Arab Youth Survey (2019) findings that: (1) 31% of youth knew of someone with mental health issues, such as anxiety or depression, but 69% reported they did not know of anyone? (2) 49% of youth viewed mental illness as a normal part of life but 50% viewed it negatively?
5. How does your family and friends talk about mental health issues such as depression and anxiety?

References

Al-Krenawi A (2005). Mental health practice in Arab countries. Curr Opin Psychiatry 18:560-564

Arab Youth Survey 2019. 11th Annual ASDA'A BCW Arab Youth Survey 2019: A call for reform. ASDA/A BCW. Available via https://www.arabyouthsurvey.com/. Accessed 3 August 2020

Beck AT, Steer RA & Brown G.K (1996). Manual for Beck Depression Inventory-II. San Antonio, TX: Psychological Corporation

Corrigan PW, Larson, JE, Rusch N (2009). Self-stigma and the "why try" effect: Impact on life goals and evidence-based practices. World Psychiatry 8(2):75-81

Corrigan PW, Rao D (2012). On the self-stigma of mental illness: Stages, disclosure, and strategies for change. Can J Psychiatry 57(8): 464-469. doi: https://doi.org/10.1177/070674371205700804

Corrigan PW, Watson A (2002). Understanding the impact of stigma on people with mental illness. World Psychiatry 1(1):16–20

Dahlgren G, Whitehead M (1991). Policies and Strategies to Promote Social Equity in Health. Stockholm, Sweden: Institute for Futures Studies

Dardas LA, Bailey DE, Simmons LA (2016). Adolescent depression in the Arab region: A systematic literature review. Issues Mental Health Nursing 37(8):569:585. doi: https://doi.org/10.1080/01612840.2016.1177760

Dardas LA, Silva SG, van de Water B, Vance A, Smoski MJ, Noonan D, Simmons, LA (2019a). Psychosocial correlates of Jordanian adolescents' help-seeking intentions for depression: Findings from a nationally representative school survey. The Journal of School Nursing 35(2):117-127

Dardas LA, Shoqirat N, Abu-Hassan H, Shanti BF, Al-Khayat A, Allen DH, Simmons LA (2019b). Depression in Arab adolescents: A qualitative study. Journal of Psychosocial Nursing 57(10):34-43

Dardas LA, Silva SG, Smoski M, Noonan D, Simmons A (2017). Personal and Perceived Depression Stigma among Arab Adolescents: Associations with Depression Severity and Personal Characteristics. Archives of Psychiatric Nursing 31:499-506

Engel G (1977). The need for a new medical model: a challenge for biomedicine. Science 196:129-136

Garritty C, Gartlehner G, Nussbaumer-Streit B, King VJ, Hamel C, Kamel C, Affengruber L, Stevens A (2020). Cochrane rapid reviews methods group offers evidence-informed guidance to conduct rapid reviews. J Clin Epidemiol, 14(130), 13-22 doi: https://doi.org/10.1016/j.jclinepi.2020.10.007

Gearing, R.E., Schwalbe, C.S., MacKenzie, M.J., Brewer, K.B., Ibrahim, R.W., Olimat, H.S., Al-Makhamreh, S.S., Mian, I., and Al-Krenawi, A. 2013. Adaptation and translation of mental health interventions in Middle Eastern Arab countries: A systematic review of barriers to and strategies for effective treatment implementation. Int J Social Psychiatry 59(7): 671-681

Gibson K, Cartwright C (2014) Young people's experiences of mobile phone text counselling: Balancing connection and control. Children and Youth Services Review 43:96-104

Gluckman P, Hayne H (2011). Improving the Transition: Reducing Social and Psychological Morbidity During Adolescence. Office of the Prime Minister's Science Advisory Committee. New Zealand Government. Available via http://www.pmcsa.org.nz/wp-content/uploads/Improving-the-Transition-report.pdf. Accessed 4 August 2020

Goffman E (1963). Stigma: Notes on the management of spoiled identity. New York: Simon & Schuster

Hannes K (2011). Chapter 4: Critical appraisal of qualitative research. In: Noyes J, Booth A, Hannes K, Harden A, Harris J, Lewin S, Lockwood C (eds.), Supplementary Guidance for Inclusion of Qualitative Research in Cochrane Systematic Reviews of Interventions. Version 1 (updated August 2011). Cochrane Collaboration Qualitative Methods Group. Available via http://cqrmg.cochrane.org/supplemental-handbook-guidance

Larrier, YI, Allen, MD, Larrier IM (2017). The role of stigma in the global mental health crisis: A literature review. Journal of Global Engagement and Transformation, 1(1), Available via https://everypiecematters.com/jget/volume01-issue01/the-role-of-stigma-in-the-global-mental-health-crisis-a-literature-review.html

Liu J, Modrek S, Sieverding M (2017). The mental health of youth and young adults during the transition to adulthood in Egypt. Demographic Research 36:1721-175. doi: https://doi.org/10.4054/DemRes.2017.36.56

Maalouf FT, Alamiri B, Atweh SE, Becker AE, Cheour M, Darwish H, Ghandour LA, Ghuloum S, Hamze M, Karam E, Khoury B, Khoury SJ, Mokdad A, Meho LI, Okasha T, Reed GM, Sbaity E, Zeinoun P, Akl, EA (2019). Mental health research in the Arab region: Challenges and call for action. Lancet Psychiatry, 6:961–66

Malatest International (2016). Evaluation report: The youth primary mental health service. Available: https://www.health.govt.nz/system/files/documents/publications/evaluation-report-youth-primary-mental-health-service-dec16.pdf. Accessed 4 August 2020.

Marmot M, Wilkinson R (eds) (2006). Social Determinants of Health, 2nd Edition. M Oxford: Oxford University Press

Merhej, R (2019). Stigma on mental illness in the Arab world: beyond the socio-cultural barriers. Int J Human Rights Care, 12(4)285-298

Radcliffe, D. & Abuhmaid, H. 2020. Social media in the Middle East: 2019 in review. Available via https://ssrn.com/abstract=3517916 or https://doi.org/10.2139/ssrn.3517916 Accessed 4 August 2020.

Sewilam, AM, Watson, AMM, Kassem, AM, Clifton S, McDonald, MC, Lipski R, Deshpande S, Mansour H, Nimgaonkar VL (2015). Roadmap to reduce the stigma of mental illness in the Middle East. Int J Soc Psychiatry 61(2):111–120. doi:https://doi.org/10.1177/0020764014537234

Social Wellbeing Agency 2018. Prime Minister's youth mental health project. Available via https://thehub.swa.govt.nz/resources/prime-ministers-youth-mental-health-project/ Accessed 4 August 2020

Superu 2017. Improving youth mental health. What has worked, what else could be done? Superu: Social Policy Evaluation & Research Unit. Available via https://thehub.swa.govt.nz/assets/Uploads/Youth-Mental-Health-Project-Research-Summary.pdf. Accessed 4 August 2020

WHO 2020a. Adolescent mental health. Available via https://www.who.int/news-room/fact-sheets/detail/adolescent-mental-health. Accessed 3 August 2020

WHO 2020b. Child and adolescent health. Available via https://www.who.int/mental_health/maternal-child/child_adolescent/en/. Accessed 3 August 2020

WHO 2020c. Mental health: Strengthening our response. Available via https://www.who.int/news-room/fact-sheets/detail/mental-health-strengthening-our-response. Accessed 3 August 2020

Youthline 2017. Youthline: Changing Lives. Available: https://www.youthline.co.nz/. Accessed 3 August 2020

Zolezzi M Alamri M, Shaar S, Rainkie D (2018). Stigma associated with mental illness and its treatment in the Arab culture: A systematic review. Int J of Social Psychiatry 64(6):597-609

Part III
Policy and Programs Needs That Target Adolescent Mental Health in the MENA

Developing and Implementing Youth-Friendly Public Policies: *A Perspective into the Arab Region*

Jennifer Dabis and Hala Allabadi

1 Introduction

Youth is a period of transition, from childhood to adulthood in which young people start to prepare for the responsibilities ahead, including their livelihood, education, employment, and.

becoming active citizens within their communities (Casey et al. 2010). Today, youth comprise approximately a quarter of the world's population (UNFPA 2014). According to the United Nations, by 2020, 3.6 billion people in the world will be aged below 30, making this population group as the "new global power reshaping the world." Young people have great potential for economic development and the creation of economic and social prosperity; however, youth are seen as a tremendous challenge to governments around the world because they are seen as too young to make a difference and be a part of the decision-making process. Thus, they have been a neglected population in international strategies and government development plans (Youth Policy Manual for Arab Countries 2016). There is an increasing rate of young people who do not have opportunities for establishment of goals, education, employment, or engagement in communities. Globally, over 200 million adolescents are not in school, and for those who are, they are not receiving the right education and skills that they need. In the Middle East and North Africa (MENA) region, education has become a source of widespread frustration because it is not delivering

J. Dabis (✉)
Juzoor for Health and Social Development, Al-Bireh, Palestine
e-mail: jdabis@juzoor.org

H. Allabadi
Public Health and Epidemiology, An Najah National University, Nablus, Palestine
e-mail: hlabadi@juzoor.org

C. Barakat, F. Al Anouti (eds.), *Adolescent Mental Health in The Middle East and North Africa*, Global Perspectives on Health Geography,
https://doi.org/10.1007/978-3-030-91790-6_5

the skills young people need in today's world, leading them to face unemployment. There is a need to invest in quality education and improve learning for youth as education in the long run can help drive long-term economic growth, strengthen institutions, and promote social cohesion. Additionally, the living conditions of youth across societies and communities drastically vary from one individual to another, and from region to region, which creates different challenges in relation to policies that help young people to cope with their needs and expectations as well as to integrate these needs successfully in their lifestyles and living conditions (UNESCO 2019).

In the MENA region, opportunities for youth in the region are severely limited due to conflict, poverty, and various disadvantages (Fehling et al. 2016). However, over the recent years, youth have been at the forefront of the various calls witnessed in the MENA region. Continuously, the message echoed throughout the region is the urgent need to ensure opportunities for the role of youth in society and the need for public services to respond to their needs adequately (MENA-OECD 2018).

The current Arab region is home to the youngest population in the world and the largest generation the region has witnessed in 50 years, with more than 60% of the population under the age of 30 (UNDP 2016). While these trends have caused an unprecedented "youth bulge," Arab youth are also facing development challenges and are living under devastating social and political conditions (UNESCO 2020). Arab youth face high unemployment rates and undergo difficult economic conditions, even though they tend to be more educated than older ages. Unemployment rates among Arab youth is 29% versus 13% worldwide, with rates for Arab women twice that of young men, with the figures reaching 25% and 44%, respectively (World Bank 2017). Additionally, formal political participation rates among Arab youth are of the lowest worldwide, reflecting low levels of trust in political institutions and processes and in the state (UNDP 2013).

In many Arab states, the rate of jobs remains low due to the lack of government intervention to create policies and programs that will lead to an increase in employment, in addition to the mismatch between a "higher education approach" and job-market demands (Youth 2011). According to the Arab Human Development Report 2016, youth in the Arab countries have great difficulty voicing their expectations and experience less satisfaction and less control over their future than otherwise similar youth in other parts of the world. Similarly, youth are not fully involved in the design of development policies for their future. Their exclusion is not necessarily targeted at them but is linked to factors such as social class, gender, geographic location, culture and/or community. Consequently, Arab youth are demanding responsive political systems and social justice, while seeking equitable economic opportunities (Arab Human Development Report 2016).

Among the 22 countries in the Arab region, only nine (Tunisia, Egypt, Bahrain, Saudi Arabia, Oman, Qatar, Kuwait, United Arab Emirates, Morocco) have developed youth policies or are in the process of establishing them. Recent political movements and the uprising of the Arab spring have highlighted the importance of governmental policymakers to focus on developing youth policies. Nevertheless, recent unrest in the region has presented a convincing rationale for governments to continuously bring youth concerns into the focus of the national agenda. Policymakers have increasingly recognized the need to formulate national youth policies and action plans, aimed not only at fostering youth development, but also at providing them with the necessary opportunities to reach their full potentials in education, health, employment, and participation in public and political life.

Youth policy is considered a special field of policy: it is designed to improve young people's living conditions and provide equal opportunities as well as to support their possibilities to participate in economic, social, and political life, mainly in education and the labor force. It is about creating conditions for youth to learn and develop the adequate knowledge and skills to excel as well as the competencies to explore the roles they desire within their communities. Developing a youth policy will ensure the opportunities for young people to be active citizens, actors of democracy, integrated into society, and play an active role in both civil society and the labor market (Adulllt 2019).

It is important not to overlook Arab youth aspirations for participation in the decision-making process, as they are more aware of the importance of participation and its relevance to them, their societies, and their future. Calls have been initiated in the region on improving technical and vocational training but still much is needed to minimize the gap within the labor market.

This chapter review sought to identify and assess the current knowledge, existing youth policies, as well as gaps in youth policies related to the needs of youth living in the Arab region, including Palestine, Lebanon, Egypt, and Tunisia. These countries were chosen due to their central position in many of the acute issues facing vulnerable youth in the region.

2 Methodology

This literature review focuses on youth and youth policies in the MENA and Arab region. There is no specific consensus on what "youth" is; however, in this chapter, youth refers to a period of complex transitions in the life course of individuals from their childhood until they become adults. The definition of youth may vary between countries. In this chapter, it will be considered as youth between the ages of 15 and 29 years old. The second word in the phrase is policy, which is a process by which public resources are allocated to achieve political objectives. When speaking about youth policy, one needs to pay attention to what is actually meant by policy: the process of making youth policy, its focus or the policy domains covered, or its application, how it is delivered for young people. In this book chapter we will focus on what existing policies are for youth in these countries, what the gaps are in youth policies in these countries, the process of making youth policies, and which domains are covered among policies in these particular Arab countries.

2.1 Data Extraction and Search Process

This literature review used a pre-specified search method to ensure that the findings were specific to only the criteria for this book chapter. Identified articles were iteratively screened for relevance regarding youth and youth policies in the MENA and Arab region by article title, abstract, and then full article. Reference pages of included articles were also reviewed to identify additional relevant references. The search was conducted in Google Scholar, PsycINFO, UNFPA database, UNICEF database, and the UNESCO database. We also searched the databases of key development agencies including the World Bank and the Department for International Development. These databases were chosen due to their broad coverage of topics relevant to youth work and policy change. Due to time constraints, the number of potential documents screened was limited for each database prior to beginning the search. Search terms were pre-specified to ensure transparency and replicability. Search terms included: "young Arab"; "youth Arab"; "youth policy" AND "Arab"; "youth" AND "MENA"; "youth policy" AND "MENA"; "young Arab" AND "opportunities"; "youth participation" AND "Arab"; "youth voices" AND "Arab"; "youth" AND "Palestine"; "youth policy" AND "Palestine"; "youth" AND "Lebanon"; "youth policy" AND "Lebanon"; "youth" AND "Tunisia"; "youth policy" AND "Tunisia"; "youth" AND "Egypt"; "youth policy" AND "Egypt."

This process of selection of documents from abstracts, titles, and relevant references resulted in a selection of 64 articles and/or reports, which we thought were relevant to our topic. Due to restrictions in references, we narrowed down the number of documents retrieved from our search to 30, by including documents related to most recent youth and youth policy information in the Arab region, and particularly in Palestine, Lebanon, Tunisia, and Egypt.

2.2 Inclusion and Exclusion Criteria

Population: The targeted sample in our literature review was comprised of youth aged 15–29 years residing in the Arab region and in particular Palestine, Lebanon, Tunisia, and Egypt.

Articles and reports: Articles and reports chosen for inclusion in the chapter were those aimed at providing knowledge on youth and youth policies in the aforementioned countries and in the Arab region in general.

Date: Data extracted must have been published from 2000 onwards and must document information on youth and youth policies taken since 2000 to ensure more recent data and relevance to youth practices and policies conducted in our specified countries.

The following section will cover the results found on youth and youth policy in Palestine, Lebanon, Tunisia, and Egypt.

The findings of the countries discussed in this chapter are not comprehensive; however, they are a starting point for clearly defined and effective policies and interventions to help improve the lives of youth living in Arab countries. Each section will discuss the definition of youth, challenges facing youth, the importance and the gaps in youth policy, and the development of youth policy, for each of the four countries.

3 Results

3.1 Palestine

In Palestine, youth (aged 15–29) comprise 30% of the total population, while children and youth under age 29 comprise 50% of the population. These rates are estimated to continue growing at a very rapid rate (UNFPA 2017). In the face of a growing population with great potential and opportunities, youth remain disempowered, disenfranchised, and disenchanted among Palestinian communities. Since the beginning of the Arab-Israeli War, Palestine has been under occupation, drastically impacting the development of the country in all aspects. The ongoing Palestinian-Israeli conflict has inevitably influenced youth: externally through the effect of the occupation including movement restrictions and internally, by the fragmentation occurring across the country, loss of identity, and a patriarchal society. Furthermore, exposure to high rates of violence, discrimination, and humiliation can be tremendously damaging during the transition between childhood and adulthood (Floris 2019). Surrounded by a significant military presence and checkpoints, Palestinian youth are regularly subjected and exposed to harassment, beatings, and shootings on the way to and from school. As a result, the reality is that unless substantial investments are directed to under-resourced child and adolescent protection services, children and adolescents in need will be at significant risk of developing serious mental health issues (UNICEF 2019). Thereby, the ongoing political conflict is causing considerable challenges for youth to fulfil their real potential and to become active citizens.

Youth concerns and policies have been on the Palestinian agenda during recent years, aiming to target the most vulnerable youth. Palestine is currently facing key challenges in regard to empowerment of youth. However, if sustainable, long-term strategic policy plans and interventions are developed for youth, their growing potential could lead to positive growth for Palestine as a whole. Key challenges Palestinian youth are experiencing are lack of quality access to education and economic opportunity; participation in public and political life; access to quality youth-friendly healthcare, including sexual and reproductive health (SRH); and gender dynamics, while also feeling they are invaluable members of society. With the aforementioned concerns among the Palestinian society, it is critical now for the Palestinian Authority (PA) and relevant partners to invest in young people, not only for the future of Palestine, but for today.

In recent years, strategic opportunities and policies to address the key challenges faced by Palestinian youth have been on policy agendas including key focus areas such as economic empowerment (high unemployment, poor quality of education, gender-normative labor segregation); civic engagement (poor youth representation, limited decision-making power); health (limited access to holistic services, sexual and reproductive health); dependence on unhealthy coping mechanisms; and gender (patriarchy, oppression among youth and women, widespread gender-based violence must be combated, particularly with support from men). The lack of opportunities for youth combined with the youth bulge has led to staggering rates of unemployment in Palestine. In 2016, youth unemployment rates countrywide reached 39% in the West Bank and approximately 60% in the Gaza Strip. Almost 51% of university graduates are also unemployed. As a result of gender norms in the country, women are more likely to be unemployed than men as they have further difficulties to find jobs as many may be considered socially unaccepted. Similarly, women suffer from early marriage, marital and familial violence, and limited personal autonomy and are left unable to ask for help, especially due to the societal stigma around violence. Access to education in Palestine is extremely high, with 96% overall literacy and almost 100% youth literacy. Despite this, the higher the education level, the lower the employment possibilities, particularly for women due to the limited opportunities within the labor market (PCBS 2015).

On another note, adolescent and youth health concerns that require urgent attention include sexual and reproductive health; psychosocial support; the need for safe, youth-friendly health centers; and the increasing rates of unhealthy coping mechanisms including substance abuse, smoking, and consumption of alcohol (also known as risky behaviors). Although the Palestinian Ministry of Health Strategy 2014–2016 outlined one of their main objectives as including youth-friendly health services, there is a lack of focus on sexual and reproductive health and rights (SRHR)—a great oversight and one of the main gaps in Palestinian youth policies (Glick et al. 2018). It has been established that existing and future youth policies have also been focused on the most vulnerable populations in Palestinian communities including adolescent girls, refugee youth, poor youth, ex-detainees (both in Israel and Palestine), and Bedouin youth. There are still major gaps in policy and strategic plans among the most four common areas of economic empowerment, civic engagement, health, and gender. The establishment of these domains in policy reforms will empower young people in Palestine to become leaders, decision-makers, and change-makers (UNFPA 2017).

3.2 Lebanon

As defined by the Lebanese Ministry of Youth and Sports, youth in Lebanon are categorized as those between 15 and 29 years old and represent 28% of the total population. Challenges in Lebanon among youth are (1) the quality of education—with given disparities between private and public schools; (2) lack of sufficient information and services for youth on SRH, including HIV as well as the spread of substance and drug abuse; (3) low levels of political and public issues; (4) migration of young people, which represents mostly male people, and high levels of unemployment, reaching almost 66%; and (5) lack of confidence in their abilities. Although Lebanon still faces many challenges regarding youth, it is one of the very few countries which has a solid foundation and build for youth policy. Youth policy reform in Lebanon began in 2000 by a group of youth associations in collaboration with the United Nations Group for Youth and the Lebanese Ministry of Youth and Sports. Afterwards, a forum of youth policies was established in Lebanon, which caught much attention and earned a lot of satisfaction among youth and organizations and agencies in Lebanon. A document was established and distributed to all stakeholders, government agencies, NGOs, and even youth. The youth policy forum was approved by the cabinet in 2012 and was comprised of recommendations for youth policy according to five sectors: demographics and migration, education and culture, social integration and political participation, employment and economic participation, and health.

In Lebanon it has been outlined that the importance of youth policy and its impact on youth includes the following points: dealing with young people as dignified citizens; equality between young people and other citizens; limiting the client list, dependent, sectarian relationship; enhancing the participation of youth in all areas of activity in a way that speeds up the process of development reinforcing young peoples' self-confidence and self-respect; and contributing to the development of an active youth aware of their rights. Currently Lebanon also tries to work on the activation of youth participation in public life, as it is a basis for the success of youth policy in general and more specifically for the enhancement of social integration.

In Lebanon, the education sector suffers from high costs and high rates of people that did not finish high school, as well as the lack of necessary academic and life competencies, which leads to challenges to employment. As in Palestine, Lebanon faces major challenges in regard to lack of SRH in the country, and increased rates of substance/drug abuse (Sida 2012). Fortunately, with these challenges affecting Lebanese youth, in 2009, the Ministry of Youth and Sports devised a special department to deal with youth-related matters, and in 2012, Lebanon's national youth policy finally came to light. However, the endorsement of the Youth Policy Strategy has so far not translated into policy implementation, due to the impact the uprising and conflict in Syria had on political life in Lebanon. Currently, youth-related matters in Lebanon have been partially side-lined due to other political matters and burdens across the country (Fakhoury 2016).

3.3 Egypt

Youth among the Egyptian Youth Council comprise young people aged 18 to 30 years old, which differs from that of the National Democratic Party defining young people between 18 and 35 years old (Abdelhay 2019).

Youth in Egypt constitute approximately 21% of the general population with approximately 20.2 million aged between 18 and 29 years, in which 50.6% are males and 49.4% are females, according to Egypt's Central Agency for Public Mobilization and Statistics in 2018. At the national level, the Egyptian constitution of 2014 includes several articles that ensure freedom of political participation and encourages citizens to exercise this freedom. As expressed in article 65, freedom of thought and opinion is legitimate, as well as other articles stating the right to organize public meetings and demonstrations, formation of political parties, and the right to establish non-governmental organizations. The Egyptian constitution encourages political and social participation by young people, although existing articles within national constitutions actual implementation of articles are questionable (Abdelhay 2019).

Egypt's Vision 2030 has developed the first Egyptian National Strategy for Sustainable Development, which highlights the objectives to promote the inclusiveness and empowerment of young people in the country. President Abdel-Fattah Al-Sisi has also named 2016 as the official "year of the youth," and the president's office has put in place a youth leadership program—the first cohort of which includes around 500 young people. On the international level, Egypt has signed the United Nations' sustainable development goals, specifically the tenth goal on engagement of citizens of all ages in political participation. Although young people played a major role during the 2011 and 2013 revolutions, their participation in referendums and elections is low. Data from the 2014 Presidential Elections Poll indicates that the share of young people aged 18–30 who actually voted constituted only 27% of the total interviewed voters (Magued Osman 2016).

The overall policy for reproductive health with regard to youth and adolescents has been improving over the years but is still in need of additional support. Other reproductive health services are only available to married females. Harmful practices should be addressed in local institutions, schools, family centers, and from youth to elders. Despite the legal ban, the Demographic and Health Surveys in 2014 has revealed that 92% of females have had female genital mutilation (Egypt Demographic and Health Survey 2014). SRH information is not provided at schools and youth have reported their source of information is usually their peers and families. Existing donors, non-governmental organizations, and local governments are working towards improving the status of young people in Egypt. Young people in Egypt face serious issues that have direct implications on the future of the Egyptian society, unemployment and migration due to the lack of opportunities and poor living conditions (UNFPA 2020).

Education in Egypt remains a challenge due to restrictive teaching styles, the lack of encouragement for youth participation, and the act of corporal punishment. This ultimately prevents young people from developing their full potential. Many schools have poor infrastructure, lacking basic needs such as functional water and sanitation facilities. A youth survey for the Arab Human Development Report in 2002 has identified a high desire for emigration among youth in Egypt; the identified source of alienation was related to both economic and social factors (UNICEF 2014).

3.4 Tunisia

Youth in Tunisia is defined as those aged 15–19 years old. They define one of the largest population groups representing 29% of the total population and 43% of the working age group. According to a report on youth work in Tunisia in 2013, the government traditionally focuses on the age bracket 15–25 years. The report has also stressed the main challenge faced by youth as unemployment which has increased to 32% after the revolution. The report has also pointed to the fact that Tunisia has no national youth policy (Bank 2014).

Data has shown an increased number of youth participating in events or civil actions in comparison to before the revolution, with males at a higher percentage than females. The revolution stirred change in many young people who were enthusiastic about that change and wanted to be part of the events. Tunisian youth were found the most active with the Arab region in creating youth activities—mostly via youth centers—to promote healthy lifestyles and prevent risky behaviors. Nevertheless, the revolution in January 2011 pointed out large measures in regard to the social, economic, and political exclusion. One of the underlying causes of the situation is due to the lack of opportunities available to young people. Tunisia's youth unemployment rate was particularly high at 30% for those aged between 15 and 24 years. The overall unemployment rate was 14% making the ratio of youth-to-adult unemployment ratio 3:2. According to the 2005–2009 Labour Force Survey data showed that 85% of the unemployed were between 15 and 35 years of age. The survey also revealed that young people were generally less optimistic about the future than they had been in 2000. The revolution has revealed the need of youth to demand dignity, employment, freedom, and better opportunities for their employment (Bank 2012).

History has revealed that youth are keen on the government's agenda, as a declaration made in 1987 has prioritized youth "the era of change." This was followed by measures that encouraged young people to become decision-makers and to hold a seat in elected committees in national consultative bodies (Floris 2019).

Although the educational system was influenced by the French, an emphasis of the Arabic language and culture is prioritized. During 2016, the Tunisian government developed a strategic plan for the education sector, which was addressed to reduce the rates of dropouts through improving teacher training and upgrading the curricula and school infrastructure. However, SRHR is not mentioned within the school curriculum; therefore, most of the basic rights and needs of adolescents and youth are not met. Early sexual practice among young people, unemployment, social censorship, and lack of information about young people's sexuality and reproductive rights encourage risky behaviors among young people living in Tunisia (Amroussia et al. 2016).

4 Discussion

The purpose of creating youth-friendly public policies is to ensure youth have opportunities to contribute to the development of their countries. To do so, countries should address youth, schools, universities, municipalities, and governmental sectors to collectively join forces to ensure that youth are a resource to a country and not a problem.

Throughout this chapter, it has been evident that Arab countries within the MENA region tend to work sporadically on youth development and in addressing the needs of adolescents. A national strategy implemented by public authorities to provide young people with opportunities and experiences that support their successful integration into society and enable them to be responsible members and agents of change within their communities is needed. Public authorities involved in the creation of national youth-friendly public policies should include youth councils, youth NGOs, interest groups, youth groups, youth workers, youth researchers, young people, schools, teachers, employers, medical staff, social workers, faith groups, the media, and many more to ensure a comprehensive participatory approach. This is a dynamic process with systems and structures that change constantly and should be revisited (Union 2020).

Participation is a fundamental democratic right. To implement these policies, all existing barriers to youth political participation should be removed. From a more purely pragmatic perspective, if young people have the perception that formal political processes are not accessible, this can shape their attitudes for a lifetime, which leads to negative impacts on a country's political culture. All parties should be involved in developing a youth-friendly legal framework, based on youth participation and acceptance of the parties involved; this will ensure the initiation of the process to establish youth-friendly public policies.

It is important to differentiate between meaningful youth participation and tokenistic, pseudo-participation activities. It is vital that activities and processes are not just claiming to foster youth participation but are actually taking role in decision-making. "Tokenism is when young people appear to have been given a voice, but really have little or no choice about how they participate. It is participation for participation's sake or for a photo opportunity. Young participants lack knowledge and capacities and are rarely mandated by their peers" (Glick et al. 2018).

How can youth contribute? How can they support in the development of the country and how to ensure that all members of the community are benefitting? It is essential for governments to be transparent; youth have the right to be informed about the purpose, scope, and procedures of the process in developing these public policies and should be aware of the potential impact of these policies. Public policies should be respectful and rights-based; they should be approached as active agents with the right to participate and voices to be heard. To ensure that participation is not a one-time event, there is a need for development of an accountability mechanism to ensure follow-up, implementation of youth decisions, and youth constituencies. The policies that address youth participation should be youth-friendly and relevant, providing them with the opportunity to discuss and decide their priorities, methods, and techniques. The environment should be adapted to suit the needs of youth, with results-oriented projects with low access barriers, using easy language with integrating technology that addresses the techniques used by youth. In addition, policies need to be inclusive to give marginalized groups of youth equal chances in participating and to include young women, refugees, and young people with disabilities. Involvement and participation of the youth within the development of these policies should be voluntary and safe (UNDP 2013). The development of youth-friendly values and visions is required to ensure that these policies are responding to the needs of each country, ultimately ensuring social protection, health needs, and rights and promoting opportunities to youth.

Initiating coalitions on the national and international level will minimize the gaps between all parties: international organizations, non-governmental organizations, and ministries working on improving the status of youth. These coalitions will work collectively to avoid duplicity and unify terminologies and country perspectives of youth and their development. These coalitions can be the backbone for developing key strategies within the country's profile, advising and acting as the quality assurance for the implementation of youth-friendly public policies. As in Palestine, the Palestinian Adolescent Health Coalition was established in 2019, consisting of 22 member organizations from international and local non-governmental organization to UN agencies to relevant ministries working on adolescent and youth status in Palestine. It is also important to point out that at the regional level many countries alongside the MENA region are members in the International Association for Adolescent Health, which is a multidisciplinary, non-government organization that aims to improve the health, development, and well-being of 10- to 24-year-old adolescents and young adults (Samuels et al. 2017).

In conclusion, understanding the potential young people can invest in our communities and building a trustworthy relationship will foresee a better future for the MENA region, and providing opportunities for this group will endorse their belonging in their countries and only build a stronger sense of belonging and interest in strengthening their community. Nonetheless, it will enrich the country's opportunity to growth and economic development and stability when the young adults recognize the investment set out for them. This is a call to all MENA countries to involve young people and allow them to build a brighter future in dignity, health, and well-being.

Critical Thinking/Reflection Questions

1. What are the major bottlenecks facing governments to provide opportunities for youth?
2. How can governments be persuaded in ensuring holistic and youth-centered public policies? How can governments promote autonomy for youth?
3. What are the easiest way governments can respond to youth needs?
4. How can countries in the MENA region address the urgent challenges facing youth employment?

References

Amroussia N, Goicolea I, Hernandez A. Reproductive Health Policy in Tunisia: Women's Right to Reproductive Health and Gender Empowerment. Health Hum Rights. 2016 Dec;18(2):183-194. PMID: 28559685; PMCID: PMC5395000.

Abdelhay AT. Studies on youth policies in the mediterranean partner countries: Egypt 2019.

Adulllt Y. Youth Policy. 2019.

Arab Human Development Report 2016: Youth and the Prospects for Human Development in a Changing Reality, Executive Summary. United Nations Development Programme (UNDP). 2016.

Bank TW. Fact sheet: Tunisia 2014.

Bank TW. Breaking the barriers to youth inclusion: Tunisia. 2012.

Casey BJ, Duhoux S, Malter Cohen M. Adolescence: what do transmission, transition, and translation have to do with it? Neuron. 2010;67(5):749-60.

Egypt Demographic and Health Survey 2014.

Fakhoury T. Youth politics in Lebanon. A call for citizen empowerment: Policy brief. 2016.

Fehling M, Jarrah ZM, Tiernan ME, Albezreh S, VanRooyen MJ, Alhokair A, et al. Youth in crisis in the Middle East and North Africa: a systematic literature review and focused landscape analysis. East Mediterr Health J. 2016;21(12):916-30.

Floris S. Studies on youth policies in the mediterranean partner countries: Tunisia. 2019.

Glick P, Khammash U, Shaheen M, Brown R, Goutam P, Karam R, et al. Health risk behaviours of Palestinian youth: Findings from a representative survey. Eastern Mediterranean Health Journal. 2018;24:127-36.

Magued Osman HG. Towards Effective Youth Participation: Policy Brief 2016.

MENA-OECD. Youth Engagement and Empowerment: In Jordan, Morocco and Tunisia. 2018.

PCBS in Collaboration with the ILO announced the results of the Youth Transition from school to Work Survey [Internet]. Palestinian Central Bureau of Statistics. 2015.

Samuels F, Jones N, Abu Hamad B. Psychosocial support for adolescent girls in post-conflict settings: beyond a health systems approach. Health Policy Plan. 2017;32(suppl_5):v40-v51.

Sida. Youth Policy in Lebanon. 2012.

UNDP. Enhancing youth political participation throughout the electoral cycle 2013.

UNDP. Arab Human Development Report. Youth and the Prospects for Human Development in a Changing Reality. 2016.

UNESCO. Combining Data on Out-of-school Children, Completion and Learning to Offer a More Comprehensive View on SDG 4. Information Paper No. 64. 2019.

UNESCO. UNESCO rallies international organizations, civil society and private sector partners in a broad coalition to ensure #Learning Never Stops. 26 March 2020.

UNFPA. The power of 1.8 billion: Adolescents, youth and the transformation of the future. 2014.

UNFPA. Young people: Egypt 2020.

UNFPA. Youth in Palestine. Policy and program recommendations to address demographic risks and opportunities. 2017.

UNICEF and the Child Protection Working Group. A review of the humanitarian Mental Health and Psychosocial needs and gaps in the West Bank and Gaza. September 2019.

UNICEF. Egypt: Country report on out-of school children 2014.

Union E. Youth policy essentials. 2020.

Youth UN. Regional Overview: Youth in the Arab Region. 2011.

Youth policy manual for Arab countries: How to develop a national youth strategy. 2016.

Adolescent Health in Saudi Arabia: Policy Dimensions

Asharaf Abdul Salam and Mohd Fadzil Abdul Rashid

1 Introduction

Adolescence, a stage of transition from childhood to adulthood (refers to 10–19 years, as defined by the World Health Organization - WHO), characterizes physiological and psychological changes including sexual and emotional development. Adolescents, as the nation's pillar of human capitals and future leaders, require special concern in terms of creating economic and social opportunities and safe living environment to promote their healthy growth and development. Therefore, the ability of governments and private sectors to shape conditions of their lives in what can be challenging circumstances, including their health and well-being, is of upmost importance (Obermeyer et al. 2015). Indeed, the health status of adolescents is an important indicator for future health and health-care needs of the next generation (Moradi-Lakeh et al. 2016) and, thus, an investment (Al Buhairan and Olsson 2014).

As stated, Saudi Arabia is passing through a transformational phase with an ambitious vision for 2030, which is applicable to the health sector too (Al-Hanawi et al. 2019), largely due to sedentary lifestyles which have led to rising chronic diseases and consequent public health burden (Al-Hazzaa and AlMarzooqi 2018; Bajamal et al. 2017). This raises concerns in relation to diet and nutritional balance (Fatima et al. 2019). Moreover, Saudi adolescents face public health concerns of overweight and obesity that threaten life expectancy (Obermeyer et al. 2015; Al-Hazzaa et al. 2014; Al-Hayek et al. 2014) impacting mental health. However,

A. A. Salam (✉)
Center for Population Studies, King Saud University, Riyadh, Saudi Arabia
e-mail: asalam@ksu.edu.sa

M. F. A. Rashid
Universiti Teknologi MARA, Shah Alam, Malaysia

C. Barakat, F. Al Anouti (eds.), *Adolescent Mental Health in The Middle East and North Africa*, Global Perspectives on Health Geography,
https://doi.org/10.1007/978-3-030-91790-6_6

adolescents face difficulties in accessing health-care services, more so for those in low-income households and with mental or physical illnesses (Najjar et al. 2018). It was also revealed that adolescents in Saudi Arabia face mental health problems deserving attention (Abbas and Al Buhairan 2017).

In such situations, Al Buhairan et al. (2016) proposed specialized adolescent health clinics or transition clinics to address diabetes and other increasing chronic conditions that adolescents face, through a multidisciplinary care approach. This concept was brought into attention by Mahfouz et al. (2009) prior to developing a primary health-care team trained in adolescent health with an emphasis on mental health. Moyser and Amen (2020) described the low levels of mental health knowledge and readiness to exercise barriers of secondary school staff in Riyadh, Saudi Arabia.

Obviously, many researchers focus their interests on adolescent health as this need is central to global health goals (Al Buhairan and Olsson 2014), in particular Goal 3 of sustainable development goals (SDGs): good health and social well-being (United Nations 2015), from multiple angles or dimensions. It is due to the significant roles of adolescents and their rapidly increasing numbers (Patton et al. 2016); moreover, this age group is exposed to risks from various factors, leading to negative health impacts and calling for a multicomponent approach. Saudi Arabian society, in particular, is different from Western societies in many respects, as it is a country with strong Islamic values. Adolescent health would differ from that of children and adults, and thus raising a huge concern different from that of the West in terms of self-privacy and beliefs toward sexual health and behavior, therefore recommending adolescent-friendly-specific ward services (Alsubaie 2019; Al Buhairan and Olsson 2014; Al Buhairan 2010).

Compared with the rest of the world, Arab adolescents bear a higher burden of medical costs especially for transport injuries, cardiovascular and metabolic conditions, and mental and behavioral disorders (Obermeyer et al. 2015). Other studies also highlight some issues in relation to adolescent health such as lacking efforts and investments, inadequate access to health facilities and services including school health programs, etc. (Al Buhairan et al. 2015; Al Buhairan and Olsson 2014; Salam et al. 2012). Therefore, the acceleration of such efforts and appropriate intervention strategies are urgently needed to address the needs of adolescents and to support their health promotion. That is, policy dimension-based actions are really important to enforce awareness and knowledge of adolescent health and at the same time to prevent a further burden on the long run to cater to adolescents and health of the next generation. More importantly, it is a challenge to prepare adolescents as strong building blocks of society to lead the country in the near future. To the best of our knowledge, there are only few studies on these specific dimensions.

Although the Vision 2030 does not emphasize adolescent health, it details youth as vibrant and significant asset under "providing equal opportunities" guaranteeing development and deployment of their skills. As more than half of the population falls below 25 years, the demographic dividend, their entrepreneurship and enterprise opportunities receive greater attention. Thus, they are offered with funding to boost small- and medium-sized enterprises (Saudi Arabia 2013). Merged with

WHO's life course approach, Vision 2030 receives strength that adult health builds from adolescent years enabling health, happiness, and competence to contribute positively to society. This includes protection and promotion of health and prevention of diseases, developing networks with relevant sectors, and safeguarding from environmental risks of water, hygiene, and housing. Certainly, physical, social, psychological, and cognitive growth having lifelong consequences are built during youth and adolescents; thus, all behaviors related to substance use, sexual reproductive health, and nutrition are rooted across the life course (Brindis and Moore 2014; Chandra-Mouli et al. 2013). In short, the importance of promoting adolescent health, preventing risks, and detecting and treating morbidities and conditions to ensure well-being receives policy attention in Saudi Arabia.

Inspired by this, the current paper attempts to bring to light policy dimensions of adolescent health in Saudi Arabia to take into account priorities and behavioral modification efforts. Limited information is known about Saudi Arabian adolescents, particularly studies that focus on sexual development and attitudes: adolescent issues in Saudi Arabia vary from those of other countries not only in terms of general health but also of marital and reproductive health dimensions (Alsubaie 2019; Al Buhairan et al. 2015; Alquaiz et al. 2012, 2013; Tork and Alhosis 2015).

There are gaps in adolescent health-care needs in Saudi Arabia that demand building local capacity supportive to changes in the health-care system including education and reforms to serve the needs and also to address risky behaviors (Al Buhairan and Olsson 2014). Moradi-Lakeh et al. (2016) and Al Buhairan et al. (2015) studied health issues of adolescents in the Kingdom, which bring to light information related to the most preventable morbidity and mortality impacting the Kingdom's future adults. Not only the smoking habits but also the passive smoking that expose them to smoke are critical in Saudi Arabia, which leads to various health ailments (Al Agili and Park 2013; General Authority of Statistics 2020; Obermeyer et al. 2015). More importantly, this paper is timely brought out to promote the resilient and productive life of adolescents to cater to the development needs of Saudi Arabia concerning the interconnectedness between health and development.

2 Methodology

This research on adolescents in Saudi Arabia is carried out with a review of literature on topics of adolescent health and health services and special attention to adolescents' health needs, conditions, and causes specific to Saudi Arabia. It is based on English-written and peer-reviewed articles published in the last 5 years (i.e., 2015–2020) and by analyzing demographic data of national censuses (1974, 1992, 2004, and 2010) and demographic surveys (2017) published by General Authority of Statistics, Saudi Arabia. Demographic survey is one of the actual surveys carried out by using general population and housing census framework with an aim to obtain highly efficient and accurate estimates. With the random sampling technique, the survey society is divided into non-overlapping parts characterized by relative homogeneity. Thereafter, samples are drawn from each part, the primary sampling units, equally. Survey findings are interpolated with the previous census resulting to aggregating numbers.

Reviews are utilized to consolidate ideas, views, and research findings on adolescent issues in the Kingdom, as well as to back it up with salient facts and figures. Moreover, the review is synthesized toward policy dimensions concerning adolescent health priorities and behavior modifications. For this purpose, two levels of literature search were adopted: Google and Web of Science for adolescent health in general and on Saudi Arabia specifically. The search was conducted from December 2019 to May 2020. This search led to identification of around 50 studies. Studies on Saudi Arabia are mostly based on small samples from specific locations. Relevant research studies are quoted and cited. In addition, specific salient data are extracted from public sources and presented in this chapter in order to explain the dynamics of adolescent health in the Kingdom.

At the same time, responses from censuses or surveys are used to calculate proportions and rates to illustrate demographic changes and male-female differentials. Furthermore, results of a recent survey "The household/family health survey 2017," which focused on the adolescent age, are presented to understand the health concerns in the Kingdom.

3 Results

This chapter focuses on the adolescent population, the stage of transition from childhood to adulthood, a period of physiological and psychological changes, including puberty, which greatly influence adolescents' choices, decisions, and health (Alsubaie 2019), and they are as human potential characterized by dynamic brain development and with social capabilities defining trajectories into the next generation (Patton et al. 2016). Results are discussed in two main sections: adolescent demographics and adolescent health needs, conditions, and causes.

3.1 Adolescent Demographics

The adolescent population increased in the country from 1974 to 2010 and thereafter decreased as shown in Table 1. These hikes and slopes are reflective of the demographic transition as shown by age structure: a transition from broad expansive shape in 1974 to constrictive shape in 2010, and further continued constrictions till 2017. While the transitions until 2010 implied that the adolescent proportions were shrinking, this continued beyond 2010 with a decline in the adolescent proportions and growth rates.

These changes are viewed from the population structure depicted in the age pyramids for various periods, for the native population (Fig. 1). Salam and Mouselhy

Table 1 Number and percentage (of total population) of adolescents (10–19 years) by sex

	Male		Female		Total	
Year	Number	Percent	Number	Percent	Number	Percent
Total population						
1974	808,902	22.6	714,623	22.7	1,523,525	22.7
1992	1,784,059	18.8	1,763,996	23.6	3,548,055	20.9
2004	2,357,457	18.8	2,404,569	23.8	4,762,026	21.0
2010	2,767,391	17.8	2,452,848	21.0	5,220,239	19.2
2017	2,416,265	12.9	2,325,572	16.8	4,741,837	14.6
Natives						
1974	714,807	23.5	667,806	23.1	1,382,613	23.3
1992	1,574,328	25.3	1,557,636	25.6	3,131,964	25.4
2004	2,030,591	24.5	2,094,710	25.4	4,125,301	25.0
2010	2,219,882	23.2	2,062,618	22.4	4,282,500	22.8
2017	1,874,019	18.0	1,814,672	18.1	3,688,691	18.1
Expatriates						
1974	94,095	17.8	46,777	17.8	140,872	17.8
1992	209,731	6.4	206,360	15.0	416,091	9.0
2004	326,866	7.7	309,859	16.5	636,725	10.4
2010	547,509	9.2	390,230	15.6	937,739	11.1
2017	542,246	6.5	510,900	13.3	1,053,146	8.7

Fig. 1 Age pyramids of native population (Saudi Arabia) for 2010 and 2017

(2013), Salam et al. (2014), Al-Khraif et al. (2020), and Khraif et al. (2015) also explained this phenomenon of age structural transitions. That is, a change from expansive to constrictive shape of age structure gives rise to lowered childhood population, which moves further to upper levels of age pyramid, step by step.

There were 22.7 percent of adolescents of age 10–19 years (1,523,525) in 1974, which decreased to 14.6 percent in 2017 (4,741,837), showing a relative decline albeit absolute increases: higher declines in male adolescents (22.6% to 12.9%) than females (22.7% to 16.8%). This gradual relative decline is vividly seen in the native population for the same period (from 23.3% to 18.1%): equal among males (23.5% to 18.0%) and females (23.1% to 18.1%), as stated by Khraif et al. (2015). There is a decrease in the number of adolescents throughout the period from 2010 to 2017, both total and natives among males and females. While these changes, until 2020, are positive (increase) population growth, that of 2010–2017 reflects negative population growth (decrease), as shown in Table 2. The increase was higher in percentages during 1974–1992 (132.9%): higher among females (146.8%) than males (120.6%). This could be because of a longer intercensal interval (Salam et al. 2014). On the other hand, the percentage increase was subsequently relatively lower. These total figures are not reflective of the native population, which increased at a slower rate until 2010 and decreased thereafter at a higher rate for both males and females. A higher annual growth rate was also reported during 1974 and 1992, both among total and native males and females: positive until 2010 and negative thereafter.

Despite a decline in the number and percentages of native adolescents, it raises health demands, in addition to challenges in urban living, digital disruptions, etc.,

Table 2 Increase and annual growth rate in the number of adolescents (10–19 years)

	Increase						Growth rate		
	Male		Female		Total				
Year	Number	Percent	Number	Percent	Number	Percent	Male	Female	Total
Total population									
1974–1992	975,157	120.6	1,049,373	146.8	2,024,530	132.9	4.3	4.9	4.5
1992–2004	573,398	32.1	640,573	36.3	1,213,971	34.2	2.3	2.6	2.5
2004–2010	409,934	17.4	48,279	2.0	458,213	9.6	2.9	0.4	1.6
2010–2017	−351,126	−12.7	−127,276	−5.2	−478,402	−0.8	−1.9	−9.2	−1.4
Natives									
1974–1992	859,521	120.2	889,830	133.2	1,749,351	126.5	4.1	4.6	4.4
1992–2004	456,263	29.0	537,074	34.5	993,337	41.4	1.9	1.6	1.5
2004–2010	189,291	21.2	−32,092	12.5	157,199	16.9	0.5	−0.1	0.2
2010–2017	−345,863	−20.8	−247,946	−16.8	−593,809	−18.9	−2.4	−1.8	−2.1
Expatriates									
1974–1992	115,636	122.9	159,583	341.2	275,219	195.4	4.3	8.0	5.8
1992–2004	117,135	55.9	103,499	50.2	220,634	53.0	2.4	2.2	2.3
2004–2010	220,643	67.5	80,371	25.9	301,014	47.3	2.8	1.2	2.1
2010–2017	−5263	−1.0	120,670	30.9	115,407	12.3	−0.1	3.8	1.7

influencing attitudes, behaviors, aspirations, and other issues in preparation for a healthy adolescence through services and care-based policies and actions.

3.2 *Adolescent Health Needs, Conditions, and Causes*

In today's changing world, adolescents face critical health problems, such as mental health (Mahfouz et al. 2009), obesity and related chronic illnesses (Bajamal et al. 2017; Obermeyer et al. 2015), sexual behavior-related health problems (Alsubaie 2019; Kazdouh et al. 2019; Al Buhairan et al. 2015), and many others. Mental health is identified as a major public health concern that has serious implications on adolescents' well-being (Abbas and Al Buhairan 2017), due to its contributions to educational underachievement, loss of employment, and increased health-care costs (Mahfouz et al. 2009).

Table 3 shows the percent distribution of deaths by causes among adolescents in the Arab region. Top of the list includes transport/other injuries, cardiovascular and circulatory diseases, and diarrhea, respiratory disease, and meningitis as major causes of adolescent death in the Arab region (cited in Obermeyer et al. 2015). This highlights the burden of adolescent diseases (Table 3) that calls for government interventions.

Health conditions of adolescents subsequently cause death: cardiovascular and circulatory diseases are ranked third after transport injuries and unintentional injuries. Others such as diarrhea, LRI, meningitis, cancer, NTD and malaria, diabetes, and HIV and TB are linked to unhealthy food consumption and sexual behaviors (Table 4). These statistical data indicate the importance of adolescent health to be on the forefront of government initiatives, as highlighted by Salam et al. (2012).

Furthermore, the disability-adjusted life year (DALY) as per the Global Burden of Disease is also worth attending to, in particular given the disproportionately higher incidences among adolescents in the Arab region (Table 4). Besides, these shall undergo systematic calculations for its public health risk and burden. Moradi-Lakeh et al. (2016) raise this issue of risky driving behaviors of Saudi Arabian youths, which requires to be reversed during this age to reduce its burden. There are findings suggesting chronic morbidity in adolescents (Al Buhairan et al. 2015). Both these aspects are highly important for policy dimensions. While many factors contribute to the current health status of adolescents, internal personal-related factors are very important; thus, a comprehensive approach of preventing is more effective than treatments or cures.

One of the adolescent reproductive health studies held in the Kingdom from school-going girls explains their knowledge and awareness of various concepts such as menstruation and puberty, pregnancy, antenatal care, nutrition, work, checkups, contraception, and others (Table 5). Although there are satisfactory levels of knowledge and awareness in all except care during pregnancy and contraception (Tork and Alhosis 2015), there are clear needs for more education and intervention. Awareness about pubertal changes, intermarriages, and sexual terms and knowledge of STDs

Table 3 Percent distribution of deaths by causes among adolescents of age 15 and 19 years in the Arab region, 2010[a]

Causes of death[b]	Male	Female
Transport injuries	18.7	7.4
Unintentional injuries	15.8	9.6
Cardiovascular and circulatory diseases	14.3	13.7
Diarrhea, LRI, meningitis	11.7	15.7
Intentional injuries	7.3	4.7
Cancer	7.0	7.7
NTD and malaria	5.2	6.0
Diabetes and other disorders	4.6	6.7
Neurological disorders	3.7	3.4
HIV and TB	2.1	3.8
Nutritional deficiencies	1.7	2.8
Other non-communicable diseases	1.6	2.4
Digestive diseases	1.5	1.4
Chronic respiratory diseases	1.3	1.8
War and disasters	1.0	0.7
Other communicable disorders	0.8	1.4
Liver cirrhosis	0.8	1.2
Mental and behavioral disorders	0.7	0.7
Musculoskeletal disorders	0.4	1.1
Maternal disorders	0.0	8.1
Death rate by all causes per 100,000	103.3	72.2

HIV human immunodeficiency virus, *LRI* lower respiratory infections, *LCD* non-communicable diseases, *NTD* neglected tropical diseases, *TB* tuberculosis

[a]Source: Global Burden of Disease database. Cited in Obermeyer et al. 2015

[b]Causes are displayed in decreasing order, by the percentage of deaths among males in the Arab region.

are also studied in Riyadh (see Alquaiz et al. 2012). Little is known about sexual behaviors, attitudes, and associated factors among adolescents in Saudi Arabia, due not only to cultural taboos attached to open discussions but also to negative attitudes leading to risky behaviors (Alsubaie 2019). It was also investigated for the pubertal and sexual knowledge of female adolescent students in Riyadh that sexual knowledge levels are not sufficient. There are gaps in knowledge of sexual terms like homosexuals, lesbian, and masturbation. While the HIV is most known, the other sexual diseases such as syphilis, gonorrhea, and hepatitis are rarely known to students (Alquaiz et al. 2012), as seen in Table 5.

Alsubaie (2019) deciphers the possibilities of adolescent sexual behaviors of STI risk, which requires programmatic policies based at the school level: might be addressed by accounting social, cultural, and religious influences. Kazdouh et al.

Table 4 Percent distribution of disability-adjusted life years (DALYs) by causes among male and female adolescents (15–19 years) in the Arab region, 2010[a]

Causes of DALYs[b]	Male	Female
Mental and behavioral disorders	19.2	24.9
Transport injuries	10.2	3.5
Musculoskeletal disorders	9.7	10.7
Unintentional injuries	8.3	3.9
Other NCDs	7.2	8.0
Cardiovascular and other circulatory diseases	7.1	5.4
Diarrhea, LRI, meningitis	6.6	6.5
Neurological disorders	4.4	5.3
Nutritional deficiencies	4.1	7.0
Chronic respiratory diseases	4.0	4.3
NTD and malaria	3.7	3.5
Diabetes and other endocrine disorders	3.6	5.2
Intentional injuries	3.6	1.7
Cancer	3.3	2.7
HIV and TB	1.4	1.6
Neonatal disorders	1.1	0.8
Digestive diseases	1.0	0.8
Other communicable disorders	0.8	0.7
War and disasters	0.5	0.3
Liver cirrhosis	0.4	0.4
Maternal disorders	0.0	2.9
DALYs lost from all causes per 100,000	15,342.7	14,524.1

DALYs disability-adjusted life years, *HIV* human immunodeficiency virus, *LRI* lower respiratory infections, *TB* tuberculosis

[a]Source: Global Burden of Disease database. Cited in Obermeyer et al. 2015.

[b]Causes are displayed in decreasing order, by the percentage of DALYs among males in the Arab region.

(2019) also suggest similar strategies comprising of a multicomponent approach against STIs. In Saudi Arabia, despite the difficulty to openly discuss sexual issues and sexuality, it was found that adolescents had negative attitudes toward sexual activity and were involved in risky sexual behavior (Alsubaie 2019). It includes practicing masturbation daily, experiencing sexual contact before marriage, and having friends who had experienced sexual contact. The major factors that link to this are influenced by globalization and ease of access to technology and media, and social norms with unhealthy way and manner (Kazdouh et al. 2019). They added that the factors that top up to the existing factors are a lack of knowledge, an increase of sensation-seeking behaviors, lack of parental communication about sexuality,

Table 5 Knowledge and awareness on sexual and reproductive health concerns (percentages)

Items ($N = 309$)	Pretest	Posttest
Puberty and menstruation		
Menstruation is one of the pubertal changes (agree)**	90.6	95.5
Development of axillary hair is one of the pubertal changes**	92.2	95.1
Changing voice is one of the pubertal changes**	45.0	87.4
Menstruation is a monthly bleeding*	76.1	84.1
Menstruation happens in all females**	65.4	86.1
Menstruation is a natural phenomenon	94.8	95.1
Menstruation is a sign of puberty	90.6	90.6
Menstruation starts between 10 and 17 years**	44.7	72.8
The usual interval between menstrual cycles ranges from 21 to 30 days**	23.0	52.1
The source of bleeding during menstruation is the uterus**	50.8	80.9
Pregnancy and antenatal care		
The first sign of pregnancy is missed period (agree) **	62.8	83.5
Regular antenatal follow-up is essential during pregnancy**	82.8	93.2
Proper diet, rest, and mild exercise are essential during pregnancy*	83.5	91.3
Eating nutritious food is a good health practice for pregnant women	87.1	92.2
Rest and performing light work is a good health practice during pregnancy*	82.8	89.6
Regular health checkup is a good health practice during pregnancy	81.9	87.7
Not taking special diet is one of the harmful practices during pregnancy*	51.5	65.4
Hard work is one of the harmful practices during pregnancy**	61.5	85.4
Smoking is one of the harmful practices during pregnancy **	54.7	67.3
Use of medicine without medical prescriptions is one of the harmful practices during pregnancy*	54.7	67.3
Babies develop in the uterus*	84.8	91.9
Contraceptive methods		
I know only one contraceptive method*	30.1	35.3
I know two contraceptive methods*	24.9	33.7
I know more than two contraceptive methods**	35.3	52.4
I know about contraceptive pills**	36.9	66.3
I know about contraceptive injectable hormones**	14.6	45.3
I know about contraceptive IUDs**	27.2	67.6
I know about female sterilization**	10.7	31.7
I know about the natural contraceptive methods	35.6	38.5

Source: Cited in Tork and Alhosis 2015
*, ** shows Chi-square significance at $p < 0.01$ and $p < 0.05$, respectively

and lack of sex education within the school, thus promoting risky sexual behaviors in adolescents (Table 6).

In relation to diabetes, knowledge of the disease among adolescents from Saudi Arabia as it shows that their awareness and treatment adherence and the treatment barriers are importantly perceptible, giving rise to worries and functional issues (Table 7), as studied by Al Buhairan et al. (2016).

Table 6 Pubertal and sexual knowledge of female adolescent students attending public and private school in Riyadh city

Percentage of students	Total	Private schools	Public schools
Knowledge of puberty	92.1	94.0	90.8
Awareness of puberty changes before it happened to them	70.7	69.3	71.7
Know how intermarriage occurs	64.0	65.1	63.3
Knowledge about sexual terms:			
Lesbian	70.1	77.1	66.9
Masturbation*	50.2	44.6	52.8
	33.7	45.8	28.1
Knowledge about STDs:			
HIV	95.8	96.4	95.5
Syphilis	38.3	43.4	36.0
Gonorrhea	37.9	44.6	34.8
Hepatitis B	14.6	14.5	14.6

Source: Alquaiz et al. 2012
Chi-square $p < 0.05$

A study of mental health of adolescents based on physical and psychological measurements, from school students from Abha region of Saudi Arabia, reveals various manifestations – neurotic and psychotic. They include somatization, obsessive-compulsive disorders, interpersonal sensitivity, depression, anxiety, hostility, phobia, paranoia, etc. (Mahfouz et al. 2009). These symptoms of mental health problems are determined by sociodemographic factors such as age, types of school, and parental education and employment status. Such mental conditions are further explored in the Kingdom for reasons of feeling sad, hopeless, and worried, which found chronic illnesses, relationship with parents, body images and exercise frequencies (Abbas and Al Buhairan 2017), as shown in Table 8. Mental health issues of adolescents in Saudi Arabia were deserving special attention by Al Buhairan et al. (2015). Abbas and Al Buhairan (2017) suggest effective strategies including regular screening and intervention programs to better address, direct, and control early signs.

Obesity and body response-related studies revealed that nearly 40 percent of adolescents have abnormal body mass index (BMI): underweight (19.2%), overweight (14.0%), and obese (7.8%), as reported by Fatima et al. (2019). Obesity has been studied in Saudi Arabia for various regions: comparatively higher in Jeddah, Al Khobar, and Riyadh cities (Al-Hazzaa and AlMarzooqi 2018). From Table 9, it indicates that age also relatively influences obesity rates among both male and female, at the border age from adolescent to adult.

Obesity and other conditions related to unhealthy food consumption behaviors were also remarkably high among adolescents and are a major public health concern (Al-Hazzaa et al. 2014; Bajamal et al. 2017). This issue is also faced by other parts of the Arab world, where adolescents experience an ill-health due to overweight or obesity (Obermeyer et al. 2015).

Table 7 Awareness on diabetes and related conditions among adolescents in Saudi Arabia

	Teen report	Parent report for teen
HRQoL		
Total scale score**	64.78	60.33
About my diabetes	61.98	61.43
Treatment barriers	62.79	61.93
Treatment adherence**	72.81	65.46
Worry*	55.42	50.31
Communication**	70.72	62.58
Family impact scores reported by parents of teens with T1DM**		
Total scale score	67.39	
Parental functioning score	66.72	
Family functional score	73.07	
Physical functioning	64.07	
Emotional functioning	59.80	
Social functioning	76.05	
Cognitive functioning	73.02	
Communication	78.43	
Worry	42.08	
Daily activities	65.05	
Family relationships	80.9	
Significant predictors of adolescents HRQoL		
Intercept**		
Gender**		

Source: Al Buhairan et al. 2016
* Significant at 95% CI ** $p < 0.01$

Adolescents in Saudi Arabia are exposed to a number of health risk behaviors, as shown in Table 10 (Moradi-Lakeh et al. 2016). These include overweight, smoking, hookah use, not consuming fruits and vegetables, inactivity, high BP, and others including mobile use while driving. Al Buhairan et al. (2015) also studied the health impacts of these risky behaviors. They included nutrition deficiencies, traffic safety, violence, tobacco and substance abuse, physical and mental health complaints, etc. (Table 10). Nutrition components and activities of adolescents in Saudi Arabia have been emphasized by Al-Hazzaa et al. 2014; Bajamal et al. (2017); and Al-Hazzaa and AlMarzooqi 2018. Fatima et al. (2019) argued for recognizing forms of nutritional problems leading to underweight and overweight to create preventive measures in Saudi Arabia.

A recent survey carried out by the General Authority of Statistics of Saudi Arabia in 2017 highlights risky behavior and health impacts, such as smoking, health assessments, prevalence of chronic diseases, dental carries, and victims of accidents and physical injuries, and also observance of periodic medical examination. The proportion of adolescents who smoke daily is around 2.1 percent (2.3 percent in

Table 8 Prevalence of mental symptoms among adolescent secondary school students in Abha, Saudi Arabia

	Boys	Girls	Overall
Mental symptom			
Somatization	13.2	14.2	13.5
Obsessive-compulsive	15.8	12.2*	14.5
Interpersonal sensitivity	15.2	13.8	14.7
Depression	13.1	13.9	13.4
Anxiety	14.3	14.2	14.2
Hostility	14.0	12.9	13.6
Phobic anxiety	17.9	16.4	17.3
Paranoid ideation	14.0	13.8	13.9
Psychoticism	13.4	14.8	13.9
Positive global indices of distress*			
Global severity index	14.5	16.3	15.1
Positive symptoms distress index	13.7	13.7	13.7
Positive symptoms total	15.5	14.0	14.9
Significant sociodemographic risk factors determining mental symptoms			
Somatization	Rank among brothers and sisters (eldest and youngest versus middle)		
Obsessive-compulsive	School type (Islamic versus general)		
Interpersonal sensitivity	Father education (illiterate versus educated)		
Phobic anxiety	Mother occupation (not working versus working)		
Psychoticism	Rank among brothers and sisters (eldest and youngest versus middle)		

Source: Mahfouz et al. 2009
*$p < 0.05$

Table 9 Obesity among Saudi adolescents

Overweight or obese (%)	Male	Female**
City		
Al Khobar	42.8	33.1
Jeddah	46.1	34.8
Riyadh	41.2	35.8
Total	43.6	34.8
Age		
15	45.6	34.3
16	45.4	36.8
17	43.1	30.4
18	39.9	38.7
19	46.0	42.5

Source: Al-Hazzaa and AlMarzooqi 2018
**Chi-square $p < 0.001$
Chi-square test for body mass index and waist-to-height ratio for adolescents are significant for both males and females

Table 10 Health risk behaviors among adolescents in Saudi Arabia

	Total	Male	Female
Dietary behaviors			
Breakfast intake (sometimes/daily)	54.8	62.3	46.3
Fruit intake (> – 1 serving)	38.1	43,6	31.8
Vegetable intake (> – 1 serving)	54.3	55.7	52.8
Carbonated beverage consumption (> – 2 drinks)	37.5	43.9	30.4
Energy drink consumption (> – 1 drink)	21.8	25.5	17.7
Activity			
Physical exercise (daily)	13.7	19.0	7.7
Television viewing (> – 2 hours/day)	42.4	40.4	44.7
Video game playing (yes)	55.6	68.0	41.6
Internet use (> – 2 hours/day)	30.1	26.0	34.6
Cellular phone (> – 1 hour/day)	14.8	13.2	16.6
Traffic safety			
Seatbelt using (sometimes/always)	13.8	17.0	10.2
Car taking without permission (yes)	17.9	28.6	5.9
Bullying and violence			
Exposure to bullying[a]	25.0	27.1	22.7
Exposure to violence at school[b]	20.8	28.9	11.7
Exposure to violence in community[b]	19.7	22.9	16.1
Tobacco and substance (ever use)			
Cigarette smoking	16.2	22.1	9.6
Shisha smoking	10.5	13.5	7.1
Solvent sniffing	16.2	11.5	21.4
Prescription medication use for nonmedical purpose	7.2	6.0	8.5
Alcohol consumption	1.4	2.1	0.7
Stimulant use	1.5	1.6	1.4
Marijuana use	1.0	1.6	0.4
Self-reported health status (chronic illnesses)			
Bronchial asthma	8.4	10.8	5.8
Allergies (not asthma)	4.9	4.2	5.6
Hematological disorder	3.7	3.1	4.5
Skin disorders	3.6	3.1	4.2
Musculoskeletal	1.5	1.5	1.4
Genitourinary	1.2	0.9	1.4
Diabetes	0.7	0.9	0.6
Others	4.6	5.0	4.2
Mental health			
Sadness/depression	14.3	10.1	19.0
Anxiety	6.7	4.6	9.1
Measured indicators of health status (BMI)			
Underweight	152	17.2	13.0
Healthy weight	54.8	48.8	61.5
Overweight	14.1	13.9	14.5
Obese	15.9	20.2	11.0

Source: Al Buhairan et al. 2015
[a]During the last 30 days
[b]During the past 12 months

natives and 1.5 percent in expatriates); around 98.4 percent of natives assessed themselves as in good health (98.5% in males and 98.3% in females); chronic disease prevalence varies from hypertension (0.11%), diabetes (0.54%), cancer (0.03%), and cardiovascular disease (0.13%). Dental carries are more frequent among adolescents with 17.6 percent having natural teeth loss (16.6% in males and 18.7% in females). While the proportion following regular medical checkups is lesser, those victimizing accidents are high and increasing.

A study of health risk behaviors by Al-Buhairan et al. (2015) found that many contributing factors influence obesity and other related chronic illnesses. These are classified into dietary behaviors, physical activities, tobacco and substance, and others. With those factors, the uneven food consumption-physical activity practices, many of them were engaged with obesity and unhealthy psychophysical status such as sadness/depression, bronchial asthma, and anxiety. Moradi-Lakeh et al. (2016) also studied health risk behaviors such as smoking, hookah use, not consuming fruits and vegetables, inactivity, high BP, and others including mobile use while driving.

The above discussions illustrate critical health problems of adolescents, such as mental health (Mahfouz et al. 2009), obesity and related chronic illnesses (Bajamal et al. 2017; Obermeyer et al. 2015), sexual behavior health problems (Alsubaie 2019; Kazdouh et al. 2019; Al Buhairan et al. 2015), and many others. Mental health conditions are identified as a major public health concern that has serious implications on adolescent well-being (Abbas and Al Buhairan 2017), due to the impact on educational underachievement, loss of employment, and increased health-care costs (Mahfouz et al. 2009). Obesity and other conditions related to unhealthy food consumption behaviors were also, remarkably, high and are major public health concerns (Al-Hazzaa et al. 2014; Bajamal et al. 2017). These issues are also similar in other parts of the Arab world, where adolescents experience ill-health due to overweight or obesity (Obermeyer et al. 2015). In countries such as Saudi Arabia, where it is difficult to openly discuss sexual issues and sexuality, adolescents often have negative attitudes toward sexual activity and were involved in risky sexual behavior (Alsubaie 2019).

This suggests that health-care providers play a major role in addressing and providing the necessary health-care services for adolescents (Patton et al. 2016) as their needs are different from that of children or adults: thus proposing an adolescent-specific hospital ward (Al Buhairan and Olsson 2014). Before a detailed discussion on this, there is a need to understand the factors linking to those abovementioned health problems. As mentioned by Alsubaie (2019) and Abbas and Al Buhairan (2017), the main contributing factors of mental health are caused by poverty, poor household and social disadvantage, poor relationship with parents, and negative body image associated with feeling so sad or worried, subsequently increasing the risk of exposure to adversities such as scarcity of food, poor nutrition, violence, inadequate education, and living in a neighborhood characterized by absence of social networks. Moreover, obesity and related chronic illnesses are caused by inadequate physical activity, poor diets (under- and overnutrition), tobacco use, smoke cigarettes or shisha, violence, etc. (Fatima et al. 2019; Bajamal et al. 2017;

Table 11 Health risk factors of 15–19-year-olds in Saudi Arabia

Factors	Total	Factors	Total
BMI (%)		Smoking (%)	
25–29.9 kg/m^2 30+ kg/m^2	20.9 11.6	Former smoker Current smoker	2.1 3.5
Hookah (%)		Fruits and vegetables (%)	
Daily users	1.5	No servings	39.8
Activity (%)		Blood pressure (%)	
None + insufficient	54.1	Prehypertension Stage 1 hypertension Stage 2 hypertension	20.8 1.68 0.43
Daily sitting time (hour)	4.5	Weekly TV and computer (hour)	7.6
Days eating fast food (per week)	2.2	Soda or pop (numbers per week)	2.3
Days eating processed meat (per week)	0.6	Processed meat (serving per day)	1.6
Days eating other processed food (per week)	1.4	Other processed food (servings per day)	1.5
Use handset cell phone while driving (%)		Use hands-free cell phone while driving (%)	
Sometimes Always	70.1 8.6	Sometimes Always	23.4 3.9
Follows speed limit (%) Never	24.1	Use seatbelt while driving (%) Never	60.6
Follow routine health checkups within 12 months (%)	7.3	Hospital use within 12 months (%)	46.7
Outpatient clinic use within 12 months (%)	46.1	Physician office visit within 12 months (%)	42.8
Oral examination within 12 months (%)	9.1		

Source: Moradi-Lakeh et al. 2016

Moradi-Lakeh et al. 2016; Obermeyer et al. 2015; Al Buhairan et al. 2015). Also, the factors of disparities in access to health care and services of adolescents also might contribute to the health issues among the adolescent population (Najjar et al. 2018). Table 11 shows some of the relevant indicators specific to adolescents in Saudi Arabia.

3.3 *Consolidations Toward Policy-Directive Dimensions*

Health needs of adolescents should be properly addressed by policy. This section will examine these issues from three main policy directive dimensions, namely: (i) awareness and knowledge of adolescents (Alsubaie 2019; Kazdouh et al. 2019; Alquaiz et al. 2012; Mahfouz et al. 2009); (ii) caring and paying attention to their needs, including listening and effective communication and interest in working with adolescent (Al Buhairan et al. 2015); and (iii) delivery of health systems through health care and services and physicians (Fatima et al. 2019; Al-Hanawi et al. 2019; Al-Hazzaa and AlMarzooqi 2018; Abbas and Al Buhairan 2017; Al Buhairan et al.

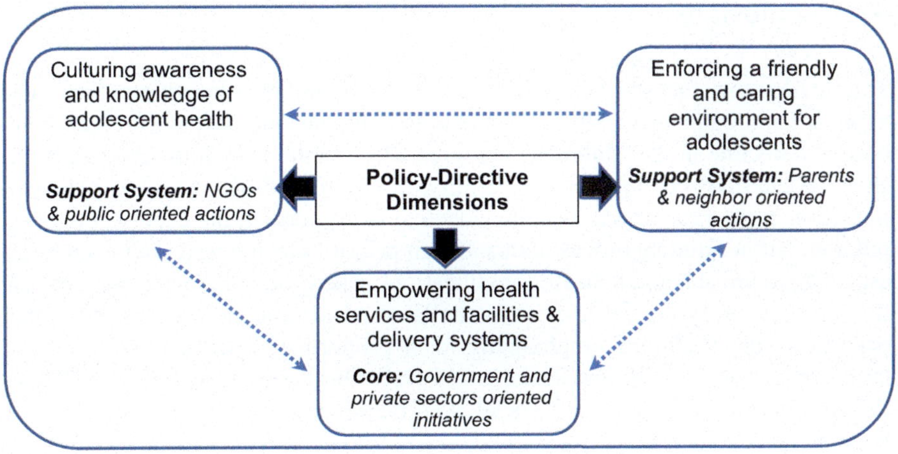

Fig. 2 Health policy-directive dimensions of adolescents in Saudi Arabia

2015; Al-Hazzaa et al. 2014). These three dimensions are a new paradigm in adolescent health in a transformational era of Vision 2030 emphasizing priorities and behavior modifications (Fig. 2).

In addressing this dynamic and complex web of adolescents' related health and behaviors, Fig. 2 depicts a new shift paradigm in health policy of adolescents concerning three directive dimensions, namely: (i) empowering health services and facilities and delivering systems as core initiative, (ii) enforcing a friendly and caring environment for adolescents, and (iii) culturing awareness and knowledge of adolescent health as support systems to the holistic health surveillance. Government and private sector serve as key players in this health policy endeavor in order to support individuals and societies. The following strategies incorporating factors linked to adolescent health problems may be adopted: (a) primary prevention through promotion of a healthful diet and active lifestyle from pre-school days; (b) sexual awareness and education; (c) engaging parents in childhood parenting programs; (d) encouragement of adolescent health research through developments such as health surveillance, nutritional database, diagnostic tools for depression and anxiety, etc.; (e) greater emphasis on challenges and concerns of adolescent age group supports with social and digital media; and (f) widening access to health-care facilities and services with fully trained physicians. Al Buhairan et al. (2015) suggest that adolescent health care be transferred from pediatrics to adult care between the ages of 15–21 years and should be tailored to meet the specific patient needs and/or combination of pediatrics and adult care. Moradi-Lakeh et al. (2016) call for a focused program for youth health (covering adolescents) in the Kingdom, placing prevention at the forefront.

4 Conclusions

This review that focused on adolescents in Saudi Arabia demonstrates two main aspects: demographics health needs and policy dimensions. Statistics reveal a proportional decline in the adolescent sub-population in the Kingdom. Such a decline shall have serious connotations on fertility, family formation, and labor force participation in the near future. Despite a decline in proportion, the adolescents as the pillar of nation-building process demand their priorities addressed comprehensively, especially on health and behavior modifications. Therefore, measures and intervention strategies based on policies are essential to protect and preserve social values, behaviors, and traditions emphasizing ideal proportions, incipient declines, and more importantly, health and social well-being. During the COVID-19 outbreaks, the importance of health and social well-being and security is crucial to all populations.

This chapter presented a range of research findings on adolescent health in the Kingdom, which focused on mental health, obesity, chronic illnesses, and reproductive health covering morbidity and mortality. There are also studies focusing on nutrition, physical activity, and health-seeking behaviors. Research efforts focus on linking pathologies to exposure factors, lifestyles, and socioeconomic profiles.

Worthy to note, this chapter has proposed a new paradigm shift in adolescent health policy concerning three directive dimensions, namely: (a) empowering health services, facilities, and delivery systems as a core initiative, (b) enforcing a friendly and caring environment for adolescents, and (c) culturing awareness and knowledge of adolescent health as support systems to the holistic health surveillance. These suggestions have the capacity to inform existing health interventions and precautions and leads to effective implementation of health programs.

As the current study based its data from censuses and small location-specific studies, further extension of this effort is recommended through primary data at the national level to explore needs, demands, and expectations of adolescent health. Nationwide large-scale studies on adolescent health espouse special attention and provisions to promote health status and strengthen health-care surveillance of adolescents. Such evidence-informed policy design dimensions have the capacity to lead to behavioral modification strategies to build a healthier generation, thus fulfilling the vision of undertaking the important responsibilities of bringing prosperity for the future generation.

Reflection Questions

How is adolescent health linked to adolescent mental health?

How can the suggested policy-directive dimensions model be implemented to target adolescent mental health?

How can primary prevention through "the promotion of a healthful diet and active lifestyle" impact adolescent mental health?

Identify and discuss gaps in research on adolescent mental health.

How are issues related to adolescent health in Saudi Arabia similar or different from other MENA countries?

Questions Addressed

1. Defining adolescent to address their health concerns within a separate specialty
2. Priority health issues to be addressed within the public health system
3. Aligning adolescent health within the purview of Vision 2030 of Saudi Arabia
4. Lessons for neighboring Arabian Gulf countries from adolescent health of Saudi Arabia
5. Path of Arabian Gulf countries in achieving international and United Nations mandate on adolescent health

References

Abbas OA and Al Buhairan F. (2017). Predictors of adolescents' mental health problems in Saudi Arabia: findings from the Jeeluna national study. Child and adolescent psychiatry and mental health; 11(Article Number 52): 1–7.

Al Agili DE; Park HK. (2013). Oral health status of male adolescent smokeless tobacco users in Saudi Arabia. Eastern Mediterranean Health Journal; 19(8): 711–719

Al Buhairan F (2010). Health care providers' opinions of adolescent health care in Saudi Arabia. Journal of adolescent health; 46(2):11

Al Buhairan F, Nasim M, Al Otaibi A, Shaheen NA, Al Jaser S, Al Alwan I. (2016). Health related quality of life and family impact of type 1 diabetes among adolescents in Saudi Arabia. Diabetes research and clinical practice; 114(1):173–179

Al Buhairan F, Olsson T. (2014). Advancing adolescent health and health services in Saudi Arabia: exploring health-care providers' training, interest, and perceptions of the health-care needs of young people. Advances in Medical Education and Practice; 5(September): 281–287

Al Buhairan, FS. Tamim, H, Al Dubayee, M. Al Dhukair, S, Al Shehri, S., Tamimi W, et al. (2015). Time for an Adolescent Health Surveillance System in Saudi Arabia: Findings from "Jeeluna"; Journal of Adolescent Health 57(3): 263–269

Al-Hanawi MK, Khan SA and Al-Borie HM. (2019). Healthcare human resource development in Saudi Arabia: emerging challenges and opportunities—a critical review. Public Health Reviews; 40(1):1–16

Al-Hayek, AA; Robert, AA.; Abbas, HM.; et al. (2014). Assessment of health-related quality of life among adolescents with type 1 diabetes mellitus in Saudi Arabia. Saudi medical journal; 35(7): 712–717

Al-Hazzaa HM and AlMarzooqi MA. (2018). Descriptive Analysis of Physical Activity Initiatives for Health Promotion in Saudi Arabia. Frontiers in Public Health; 6(Article 329):1-10

Al-Hazzaa HM, Abahussain NA, Al-Sobayel HI, Qahwaji DM, et al. (2014). Prevalence of Overweight, Obesity, and Abdominal Obesity among Urban Saudi Adolescents: Gender and Regional Variations. Journal of health and population nutrition; 32(4):634–645

Alquaiz AM, Almuneef MA, Minhas HR. (2012). Knowledge, attitudes, and resources of sex education among female adolescents in public and private schools in Central Saudi Arabia. Saudi Medical Journal; 33(9): 1001–1009

Alquaiz AM, Kazi A and Muneef MA. (2013). Determinants of sexual health knowledge in adolescent girls in schools of Riyadh-Saudi Arabia: a cross sectional study. BMC Women's Health; 13 (Article 19)

Alsubaie ASR. (2019). Exploring Sexual Behaviour and Associated Factors among Adolescents in Saudi Arabia: A Call to End Ignorance. Journal of Epidemiology and Global Health; Vol. 9(1):76–80

Bajamal E, Robbins LB, Ling J, Smith B, Pfeiffer KA, Sharma D. (2017). Physical Activity Among Female Adolescents in Jeddah, Saudi Arabia: A Health Promotion Model-Based Path Analysis. Nursing Research; 66 (6): 473–482
Brindis CD and Moore K. 2014. Improving adolescent health policy: incorporating a framework for assessing state-level policies. Annual review of public health; 35 (1):343–361
Chandra-Mouli V, Bloem P, Ferguson J. 2013. The World Health Organization's work on adolescent sexual and reproductive health. Bundesgesundheitsbl; 56(2):256–261
Fatima W, Alqhatani NSM and Ahmad NM. (2019). Assessment of Nutritional Status and its Related Factors among Female Adolescent Girls: A School based Study in Arar city, Kingdom of Saudi Arabia. International Journal of Medical Research & Health Sciences; 8(2):133–144
General Authority of Statistics, Various years. Various censuses, demographic surveys, national family/household health survey www.stats.gov.sa accessed during 2020 March–June.
Kazdouh HE, El-Ammari A, Bouftini S, Fakir SE and Achhab YE. (2019). Perceptions and intervention preferences of Moroccan adolescents, parents, and teachers regarding risks and protective factors for risky sexual behaviors leading to sexually transmitted infections in adolescents: qualitative findings. Reproductive Health; 16(1):138
Khraif, RM, Salam AA, Elsegaey, I. et al., (2015). Changing Age Structures and Ageing Scenario of the Arab World. Social Indicators Research; 121 (3):763–785
Al-Khraif R, Salam AA, Rashid, MFA. 2020. Family Demographic Transition in Saudi Arabia: Emerging Issues and Concerns. SAGE Open; 10(1):1–13
Mahfouz AA., Abdelmoneim, I, Al-Gelban, KS., et al. 2009. Adolescents' mental health in Abha city, Southwestern Saudi Arabia. Int J Psychiatry Med; 39(2):169–177
Moradi-Lakeh M, ElBcheraoui C, Tuffaha M, et al. 2016. The health of Saudi youths: current challenges and future opportunities. BMC Fam Pract; 17(Article No. 26):1-9
Moyser M; Amen B. 2020. Assessment of knowledge, readiness and barriers, female secondary school teachers and staff regarding adolescent mental health in Riyadh, Saudi Arabia, School mental health; 12(3): 650–659
Najjar S, Saadeh M, Al Buhairan F. 2018. Determinants of adolescents' perceptions on access to health care services in the Kingdom of Saudi Arabia: Jeeluna national survey findings. King Abdullah International Medical Research Center, Riyadh
Obermeyer CM, Bott S, and Sassine AJ. (2015). Arab Adolescents: Health, Gender, and Social Context J Adolesc Health 57 (2015) 252–262
Patton GC, Sawyer SM, Santelli JS, et al. (2016). Our future: a Lancet commission on adolescent health and wellbeing. Lancet 387(10036):2423–2478. doi:https://doi.org/10.1016/S0140-6736(16)00579-1
Salam AA and Mouselhy MA. 2013. Ageing in Saudi Arabia: Impact of demographic transition. Bold; 24(1):13–26
Salam AA, Elsegaey I, Khraif R, et al. 2014. Population distribution and household conditions in Saudi Arabia: reflections from the 2010 Census. Springerplus; 3(1):1–13
Salam AA, Alshekteria AA, Mohammed HAA, et al. (2012). Physical, mental, emotional and social health status of adolescents and youths in Benghazi, Libya, Eastern Mediterranean Health Journal; 18(6): 586–597
Saudi Arabia. (2013). Vision 2030: The Kingdom of Saudi Arabia. The Council of Economic and Development Affairs
Tork, HMM & Alhosis KF. (2015). Effects of reproductive health education on knowledge and attitudes among female adolescents in Saudi Arabia. J Nurs Res; 23(3), 236–242
United Nations. (2015). Sustainable Development Goals (SDGs) Knowledge Platform, Transforming Our World: The 2030 Agenda for Sustainable Development. United Nations
World Bank. Palestine's Economic Outlook - October 2017. West Bank and Gaza. 2017 https://www.worldbank.org/en/country/westbankandgaza/publication/palestine-s-economic-outlook-october-2017.

Health Policies of Adolescents in the Middle East and North Africa: Past Experiences, Current Scene, and Future Strategic Directions

Mohammed AlKhaldi, Siwaar Abouhala, Fadwa Alhalaiqa, Aya Ibrahim, Rima A. Kashash, and Akram Abusalah

Abbreviations

AA-HA	Global Accelerated Action for the Health of Adolescents
ADHD	Attention-Deficit/Hyperactivity Disorder
COVID-19	Coronavirus Disease 2019
CRC/C/GC	Convention on the Rights of the Child, the General Comment

M. AlKhaldi (✉)
Faculty of Medicine, School of Physical and Occupational Therapy, McGill University, Montreal, QC, Canada

Council on Health Research for Development, COHRED, Genève, Switzerland

University of Basel, Swiss Tropical and Public Health Institute, Department of Public Health, the unit of Health Systems and Policies, Basel, Switzerland

An-Najah National University, Faculty of Medicine and Health Sciences, Nablus, Palestine

S. Abouhala
Advancing Arab American Health Network and Allies (AAAHNA) Research Group, Tufts University, Boston, MA, USA
e-mail: siwaar.abouhala@tufts.edu

F. Alhalaiqa
Faculty of Nursing, Philadelphia University, Amman, Jordan
e-mail: fhalaiqa@philadelphia.edu.jo

A. Ibrahim
United Nations Economic and Social Commission for Western Asia, Beirut, Lebanon
e-mail: aya.ibrahim@un.org

R. A. Kashash
Palestine Red Crescent Society/Lebanon Branch, Akka Center, Ghobeiry, Lebanon

A. Abusalah
Nursing Sciences, Palestine College of Nursing, Gaza Strip, Palestine

C. Barakat, F. Al Anouti (eds.), *Adolescent Mental Health in The Middle East and North Africa*, Global Perspectives on Health Geography,
https://doi.org/10.1007/978-3-030-91790-6_7

HIV	Human Immunodeficiency Virus
IATTTYP	Inter-Agency Technical Task Team on Young People
ICU	Intensive Care Units
IDP	Internally Displaced People
KSA	Kingdom of Saudi Arabia
MDGs	Millennium Development Goals
MENA	Middle East and North Africa
MoH	Ministry of Health
MoPH	Ministry of Public Health
NCLW	National Commission for Lebanese Women
NGOs	Non-governmental Organizations
OCHA	Office for the Coordination of Humanitarian Affairs
OECD	Organisation for Economic Co-operation and Development
PCBS	Palestinian Central Bureau of Statistics
PMRS	Palestinian Medical Relief Society
RMNCAH	Reproductive, Maternal, Neonatal, Child, and Adolescent Health
RSV	Respiratory Syncytial Virus
SARS-CoV	Severe Acute Respiratory Syndrome-Associated Coronavirus
SDGs	Sustainable Development Goals
SRH	Sexual and Reproductive Health
SRHR	Sexual and Reproductive Health Rights
UN	United Nations
UNDG	United Nations Development Group
UNESCWA	United Nations Economic and Social Commission for Western Asia
UNFPA	United Nations Fund for Population Activities
UNHCR	United Nations High Commissioner for Refugees
UNICEF	United Nations International Children's Emergency Fund
UNPD	UN Population Division
UNRWA	United Nations Relief and Works Agency
UNSD	UN Statistics Division
WHO	World Health Organization

1 Part 1: The Landscape of Health Policies and Stakeholders of Adolescents in the MENA Region and Evidence from the Ground

1.1 Introduction

Around 1.2 billion of the entire global population is adolescent, and this number is expected to increase through 2050 particularly in low- and middle-income (LMC) countries (World Health Organization WHO 2018). There is a need to maintain adolescent health and improve their well-being throughout the ongoing COVID-19 pandemic and in the future, both of which may inevitably involve the systemic reduction of poverty and gender-based discrimination. The maintenance of health can be achieved by designing appropriate health policies; however, most global health policies focus on young children and adults. The WHO defines the word "adolescents" as individuals in the 10-to-19-year age group and the word "youth" as the 15-to-24-year age group. While the term "young people" covers the age range 10-to-24 years (World Health Organization n.d.), the United Nations also defines "children" as those persons up to the age of 18. It is also important to distinguish between teenagers—13-to-19-year-old individuals—and young adults, 20-to-24-year-old individuals, since the sociological, psychological, and health problems they face may differ (UN n.d).

In 1989, the Convention on the Rights of the Child was adopted to protect the civil, political, economic, social, health, and cultural rights of all children up to 18 years of age; however, it did not explicitly include adolescents. In 2016, the General Comment No. 20 (Convention on the Rights of the Child, the General Comment (CRC/C/GC/20)) on adolescent was issued because the RCR recognized that their development and capacities vary greatly from younger children. The General Comment No. 20 focuses on the recognition and respect of adolescents, the empowerment and participation of this age group in their community as active citizens, and the promotion of their health, well-being, and development. It also highlights a commitment to the promotion, protection, and fulfillment of adolescent human rights without discrimination (WHO, UN Committee on the Rights of Child 2016).

The WHO has formulated many health-related strategies whose implementation is recommended to Member States in order to maintain adequate worldwide health and well-being. Based on these strategies, several health resolutions were adopted and signed by all Member States who were required to design their national health policies (WHO 2020). As a result of these initiatives, various health policies have been implemented in many MENA nations. The agreed-upon UN resolutions include universal health coverage, disaster planning, prevention and controlling of communicable and noncommunicable diseases, and health promotion through addressing disease risk factors and promoting a healthy lifestyle, such as tobacco use, substance abuse, maternal and child health, mental health, violence, and transmitted diseases (WHO Regional Office for the Eastern Mediterranean 2019).

The WHO's Global Accelerated Action for the Health of Adolescents (AA-HA) suggests that policymakers prioritize, plan, implement, monitor, and evaluate health programs of adolescents. Adolescent health needs, disease burden, and risk factors vary across their developmental span and among the countries (WHO 2017). Since the formation of the 2030 Agenda of Sustainable Development Goals (SDGs), the WHO recommends the inclusion of adolescents in universal healthcare coverage. Globally, the implementation and maintenance of health policies related to injury prevention, mental health and well-being, maternal health, substance abuse, communicable disease, and oral health can all serve as effective measures in enhancing adolescent health (WHO 2018). However, because of a severe lack of resources and technical capacities—e.g., limited funding and an absence of regional or national databases for the monitoring and evaluation of adolescent health—the impactful application of these health policies is restrained by many challenges (Patton et al. 2016; WHO 2018).

In the MENA region, adolescents are considered a major source of socioeconomical power and educational development. They are estimated to be about one-third of the region's total population, with an expected increasing trend from 2018 to 2040 (United Nations International Children's Emergency Fund (UNICEF 2018). It is apparent that adolescents may hold major roles in contributing to social systems in the region.

To overcome common obstacles to accessing quality care and living a healthy, safe, and fulfilling live, a regional framework of Joint Strategic Actions for Young People in the Arab States/MENA Region for 2016–2017 was formulated by the United Nations Development Group (r/UNDG) Inter-Agency Technical Task Team on Young People (IATTTYP). The 2016–2017 action plan aimed to achieve SDGs among young people through empowering them to develop their capacities—physically, psychologically, spiritually, socially, emotionally, cognitively, and culturally—within a safe and supportive environment (UN IATTTYP 2018). However, implementing this plan was challenged by a lack of financial, technical, and human resources, limited multi-sectoral and integrated approaches to community development, and varying levels of engagement from different sectors, fields, and actors. Another regional action plan was introduced by the United Nations Fund for Population Activities (UNFPA) to the Arab states from 2018 to 2021. This plan is designed to provide universal access to SRH services, to end gender-based violence and harmful behavior, to create youth opportunities, and to improve population-related data systems and policy-related demographic intelligence, in both development and humanitarian settings (UNFPA 2018).

As global organizations continue to roll out plans and programs on the topic of health policy, several strategic plans related to adolescent health continue to be suggested and applied in the MENA region. As part of a larger book on adolescent health in the MENA, this chapter comprehensively explores and analyzes the adolescent health policies in the MENA region considering various interconnected themes.

1.2 A Comprehensive Overview of the Sociopolitical and Biosocial Conditions in the MENA

Adolescent Health Policies and SDGs in the MENA Region

Regional public health systems are central to the implementation of global development agendas. Across a wide spectrum of outreach and modalities, public health mechanisms both contribute to and benefit from global development. In the Millennium Development Goals (MDGs) from 2000 to 2015, three out of eight goals are directly linked to public health: MDG 4 (child mortality), MDG 5 (maternal health), and MDG 6 (combating diseases) (UN 2015). For each goal, there are specific targets and corresponding indicators to measure public health conditions. With the current SDGs (2015–2030), SDG 3 is dedicated to good health and well-being comprising 13 targets and 28 indicators to monitor progress and detect gaps in health policies (UN Statistics Division, UNSD 2020). Nonetheless, all MDGs and SDGs are interconnected with health, and health is integral to the three dimensions of sustainable development: social, economic, and environmental.

Given the importance of health for sustainable development, adolescent health is particularly important as the upcoming adult population. There are many issues under the umbrella of adolescent health that need to be tackled, but some issues are continuously considered in development agendas. For example, by considering SDG 3 further, indicator 3.7.1 refers to adolescent birth rate (aged 10–19 years) per 1000 women in that age group (UNSD 2020). This shows that SRH of adolescents has been very critical to global development throughout the MDGs and will remain critical up to 2030 in regard to the SDGs. According to the UN Population Division (UNPD) (World Population Prospects 2019), it is estimated that 62 million newborns between 2015 and 2020 are attributed to adolescent-female pregnancies worldwide. Nutrition is another major concern for adolescents' health, and the second target of SDG 2 (zero hunger) addresses the nutritional needs of adolescent girls, pregnant and lactating women, and older persons (UNSD 2020). Other conditions that impact adolescents' health such as education, gender equality, healthy environment, and economic prosperity are also addressed in the 2030 agenda for sustainable development (WHO 2015). Such conditions are variable on regional, national, and local levels, and what could be a priority issue on the global level might not be the case in a certain country. Global development agendas may not always reflect the urgent reality of adolescent health issues, and therefore, national policymaking processes are key mechanisms to communicate and improve adolescent health conditions.

Unlike the global trend that the number of persons aged 65 years or older will surpass the number of adolescents by 2050 (UNPD 2019), the MENA region's adolescent population will be at peak between 2018 and 2040 (UNICEF 2019). Children and young people account for nearly half of the MENA region's population, creating a significant opportunity for economic growth (UNICEF 2019). However, the MENA region is heterogeneous in terms of economies, resources, security, and political conditions, which determine the availability and access to health services and infrastructure for adolescents (AlBuhairan 2015). For instance, in higher-income countries such as the Kingdom of Saudi Arabia (KSA) as an example from the Gulf countries, adolescents have access to quality healthcare services and infrastructure. On the other hand, in-conflict, occupied, and politically unstable areas such as in the Syrian Arab Republic, Lebanon, and Palestine, adolescent refugees suffer from a lack of education and health services. The unsanitary and crowding conditions in refugee camps increase the risk of transmission of communicable diseases among adolescents and other age groups (AlBuhairan 2015). Moreover, cases of malnutrition and mental health disorders are prevalent among adolescent refugees due to deteriorating living conditions along with the absence of basic needs and services. Overall, the diverse conditions among countries in the MENA region dictate the disparities in adolescent health policies in this region, and those policies are key to achieve sustainable development.

MENA countries develop National Health Strategies to address gaps and set future targets and mechanisms to improve health conditions. Such strategies, yet, fail to accommodate adolescent health priorities. Theses health strategies are inappropriately programmed to support and fulfill the adolescents' needs and rights to including access to good-quality health services and education, clean air and water, adequate sanitation and good nutrition, protection from violence and discrimination, societal participation, and healthcare for chronic diseases as well as for mental illnesses and disability. The MENA region has one of the highest adolescent birth rates globally, and this rate varies within the region. According to the latest Arab Sustainable Development Report, Maghreb countries have the lowest adolescent birth rate of 11.18 per 1000 women (aged 15–19 years) compared to 18.97 in Gulf countries and the highest rate of 60.0 in Mashreq countries (United Nations Economic and Social Commission for Western Asia (UNESCWA 2020). It is argued that many countries in MENA do not sufficiently address the SRH needs of the young population due to sociocultural constraints. For example, in addressing SDG 3, Gulf countries place more efforts on SDG target 3.1 related to maternal mortality and less emphasis on SDG target 3.7 related to SRH (Daher-Nashif and Bawadi 2020). Another issue of concern is the absence of specific adolescent health strategies in the SDG reports prepared and published by Gulf countries showing that governmental health policies are not catering for this group's needs (Daher-Nashif and Bawadi 2020).

Mental health is addressed in SDG target 3.4 to reduce by one-third premature mortality from noncommunicable diseases through prevention and treatment and promote mental health and well-being (UNSD 2020). Nevertheless, indicator 3.4.2 only measures suicide mortality rate as an indication of mental health, which is not reflective of many nonfatal mental health disorders (UNESCWA 2020). War-, conflict-, and violence-affected adolescents in the MENA region have a higher risk of exposure to mental health problems. Especially in refugee and displacement camps, there is a lack of sufficient expertise and resources to diagnose such mental health disorders. Occupied Palestine has one of the highest burdens of adolescent mental health illness in the region including behavioral and emotional disorders, ranging from moderate to severe disorders (WHO 2019). Health services in the MENA region are not sufficient to accommodate or deal with the burden of mental health illness. In some Arab countries, it is estimated that there are 7.7 mental health workers and 5.1 beds per 100,000 mental health patients compared to a global average of 9.0 mental health workers and 16.4 beds (UNESCWA 2020).

While there are many efforts in the region to address adolescent health in national health policies within the context of sustainable development, there are gaps in both national and global agendas hindering progress. Further investments are required to improve and update health data infrastructure to include specific adolescent health indicators. MENA's health stakeholders should also be aware that as it is important to adopt global development agendas, it is important as well to adopt national strategies that cater for a country's adolescent population. SDGs targets and indicators are broad and are set to resolve global health issues that vary across regions (WHO EMRO 2015). Furthermore, linkages across different global agendas and within agendas help improve monitoring and planning of health strategies/policies (Kuruvilla, S., Bustreo, F. et al. 2016). For example, SDG 6 on water and sanitation is essential to prevent the spread of communicable diseases in refugee camps. The Global Strategy for Women's, Children's, and Adolescents' Health (2016–2030) is also aligned with nine SDGs, and it aims to induce transformative and inter-sectoral changes in health systems (Kuruvilla et al. 2016). Thus, improving adolescent health in the MENA region is a priority to achieve the 2030 agenda, and improving adolescent health is determined by social, economic, and environmental conditions and by how existing policies are formulated and implemented.

Globally, States are demonstrating a growing political commitment to develop comprehensive policies to better respond to the needs and aspirations of young people. Among the national health policy documents from 109 countries retrieved in 2013 from the WHO Country Planning Cycle Database, 84% of the policies included some attention to adolescents. In three-quarters of them, the focus is on SRH. Approximately one-third address tobacco and alcohol use among adolescents, and one-quarter address mental health (WHO 2014).

In MENA, although adolescents represent a large proportion of the population, their health status, healthcare needs, and adolescent health policies have not been given proportional attention. Some evidence from the Arab Gulf countries indicates that adolescents in the region demonstrate significant rates of health-compromising behaviors and conditions, such as tobacco use, lack of physical activity, poor nutrition, and obesity that contribute to noncommunicable diseases. This implies the need to enhance healthcare practice and facilities, clinical education, school health services and program, and adolescent health research to address key aspects of adolescent health and medical care (Al Makadma 2017). Therefore, it is suggested that the priority for Arab countries is to empower evidence-based advocacy for comprehensive adolescent policies and program development, investment, and implementation (Ferguson and Dick 2012). This fact is reinforced by research evidence that indicates that policy or health system research dedicated to systematically investigating adolescent health in terms of policy, healthcare system, and priorities is still lacking. This is due to various gaps such as governance, resource, capacity, prioritization, and stakeholder roles. To address these gaps, health policy and system research is imperative to inform policymakers to make the right decisions supported by evidence-based knowledge. Ultimately, this modernistic approach and adoption of this new thinking not only makes a transformational renaissance in adolescent health policies, but also in reforming other sectors, disciplines, and societies, especially in fragile and unstable countries in the region (AlKhaldi et al. 2018; Alkhaldi et al. 2020a).

Overall, Arab countries are still in need of strengthening their healthcare system capacity, advancing tobacco control legislation, promoting healthy living campaigns, and developing policies that encourage young people to adopt healthier eating habits and engage in regular exercise (Abdul Rahim et al. 2014).

Although some adolescent's health risks persist in Jordan, there is significant progress toward adolescent's health policies, which the government strives to provide young people with better access to health facilities and services, regardless of their location, sex, or social background (Organization for Economic Cooperation and Development (OECD) Development Centre 2018).

Furthermore, there is a critical gap in the existence of SRH needs of adolescents in 11 Arab countries: Algeria, Egypt, Jordan, Lebanon, Morocco, Palestine, KSA, Sudan, Syria, Tunisia, and the United Arab Emirates. Cultural taboos are a major obstacle to informed discussions about SRH issues, particularly for young people. Egypt, for example, has developed a strategy for the prevention of child marriage to support the law prohibiting early marriage (DeJong and Bashour 2016).

Stakeholders Involved and Their Contributions

Policymakers are playing a major role along with other stakeholders in issues relevant to adolescence. Policymakers face the challenge of understanding the nature of adolescents' specific needs, the implications of these for policies and programs in the health and other sectors, and how to respond to adolescents' priorities in the face of competing demands. The wide range of issues that need to be tackled range greatly: from obesity to violence and from adolescent pregnancy to depression (WHO 2016).

Countries like Jordan, Egypt, Palestine, Morocco, and Lebanon have adolescent health strategies and special bodies and stakeholders to tackle youth issues and formulate related policies. Jordan as an example is approaching toward improving young people's ability to exercise SRH and rights in development and humanitarian settings under the 2018–2022 country program (UNFPA 2020a).

There are many players contributing to the formulation of the adolescents' health policies. A systematic review study found that there are 15 participating stakeholders involved in formulating adolescent policy in Jordan: United Nations High Commissioner for Refugee (UNHCR), UNICEF, International Youth Foundation, Save the Children, Mercy Corps, Norwegian Refugee Council, International Rescue Committee, International Medical Corps, Microfund for Women, Jordanian Hashemite Fund for Human Development, Baqaa refugee camp youth center, Za'atari Syrian refugee camp, Palestinian refugee youth, Syrian refugee youth, and Jordanian youth (Fehling et al. 2015).

Moreover, Y-PEER, the Youth Peer Education Network, is established in 2013 by UNFPA in Palestine as a groundbreaking and comprehensive youth-to-youth initiative. It consists of more than 500 non-profit organizations (NGOs) and governmental institutions. Its membership includes thousands of young people working in many areas, including those involving adolescent SRH (UNFPA 2020b). On the other hand, the UNFPA attempts to meet the youth's needs through engaging them all in various stages of interventions and levels of programs, civil society organizations, and government as represented through the Lebanese Ministry of Youth. At the policy level in Lebanon, the UNFPA provides technical assistance mainly to the

National Commission for Lebanese Women (NCLW) and related NGOs to play an important role in developing the capacity to put reproductive health on the agenda in both the health and the education systems (UNFPA 2014). Other national reports revealed disparities concerning the role of civil society in SRHR, as which Tunisia, Lebanon, Egypt, and Jordan indicated that civil society has a major role in working with government and independently (DeJong and Bashour 2016). However, it was evident that the governments are the key player in formulating relevant adolescent health policies, but other key actors play a role as well.

Adolescence Health Policies: Case Studies from the Region (Sudan, Palestine, and Lebanon)

Sudan

Sudan was one of the earlier countries that took action to improve adolescent health by adopting AA-HA Guidance. Sudan did that to meet its youth needs through developing a robust national adolescent health strategy (WHO 2019). Sudanese national strategy of adolescent health and well-being established for 2018–2022 addressed many important issues. The strategy tackles the following components; (1) adolescent SRH, (2) violence against adolescents, (3) adolescent nutrition, and (4) unintentional injuries and the mental health of adolescents (WHO 2019). In Sudan, there are Reproductive, Maternal, Neonatal, Child, and Adolescent Health (RMNCAH) policies. However, the implementation of these policies is challenged by a critical shortage of human resources across the health system building blocks and levels of the health system, low coverage, low access to healthcare services, a lack of medicines and supplies, weak health leadership, and low national funding (Belaid et al. 2020). Unfortunately, there are no related statistics to determine the impact of this strategy on Sudanese adolescents. However, the Sudan situation report in 2019 reported that the number of malaria breached the epidemic threshold, humanitarian needs (nutrition, health, and protection) were high particularly in eastern Sudan, and the communicable diseases (e.g., diphtheria, dengue fever) continue to be reported across the country (Office for the Coordination of Humanitarian Affairs (OCHA) 2019). This reflects the need to design appropriate executive plan in concordance with the availability of adequate resources and leadership at high-level support to facilitate application of health policies in Sudan.

Palestine

Palestinian society has been continuously described as a young society. To support the development of effective policy and action for child and adolescent health, Palestine has stepping nascent progress toward supporting adolescent health in terms of policies and actions. The percentage of adolescents in the age group of 15–29 years in Palestine comprises 21.9% of the total population (Palestinian

Central Bureaus of Statistics (PCBS) 2018). Palestinians, in general, live under compounding challenges, instability, fragility, poor conditions, poverty, and mobility, all of which are caused by multifactorial etiology. The pandemic shows triple tragedies; Coronavirus disease (COVID-19 pandemic), ongoing occupation (politics), and intra-Palestinian divide (policies) (AlKhaldi 2020b). These conditions negatively affect both children and young people and have led to an increase in violence against children, adolescents, and girls, at the family, school, and community levels. Adolescent people in Palestine face many challenges affecting their economic, social, and health well-being. It is a unique case as it has been experiencing fragility, lack of resources, constrained freedom, and violation (UNFPA 2020b).

The key priorities for Palestinian young people are addressing the adolescent girls' vulnerability issues, gender-based violence, and proactive psychosocial support (Hamad et al. 2020). Many institutions have services not solely for adolescents. The Ministry of Health (MoH) had one youth-friendly service center in Dura, Hebron, while Al-Quds and Al-Azhar Universities in collaboration with Palestinian Medical Relief Society (PMRS) established in 2018 a "Youth Friendly Health Center" funded by Italy and technically supported by the UNFPA to provide comprehensive health services including SRH, counseling, disease, or acute illness treatment (Hamdan and Imam 2019). Also, mental health services in Palestine are provided through 16 community health clinics in West Bank, including a mental health center for children and adolescents in the North Hebron Health Directorate (MoH 2019).

Institutions involved in adolescent health in Palestine were the international non-governmental organizations including UNICEF and UNFPA. Local NGOs work on helping the increasingly poor population. Y-PEER Palestine as an example is working in the broad areas of adolescent SRH. Moreover, key governmental organizations were involved in adolescent: the MoH's strategy in primary health care is building awareness on prevention of disease transmission, the Ministry of Interior's Anti-Narcotics General Administration has education and awareness departments in all districts, the Ministry of Education administers programs and projects that work on improving schoolchildren and adolescents' well-being through youth-friendly project and utilization of the adolescents' health curriculum, and the Ministry of Social Affairs—rehabilitation centers is concerned with marginalized and unprivileged youth utilizing social care and rehabilitation centers as part of its National Social Protection Strategy in Palestine (Sayej and Qtait 2016). Therefore, all key stakeholders in Palestine including young people, community leaders, health providers, and parents should be included in the planning, implementation, and evaluation of adolescents' health policies and services based on evidence generated from research. The ongoing rapid transmission of COVID-19 in the fragile Palestinian community, especially in the Gaza Strip, constitutes a major threat, mainly on adolescence and their services. This imposes all state and non-state and community to take immediate actions and mobilize additional resources and capacity to respond effectively focusing on meeting the adolescent needs and priorities.

Lebanon

Lebanon is a demographically and politically complex country in the MENA region. Lebanon hosted around 475,075 registered Palestinian refugees (United Nations Relief and Works Agency (UNRWA) 2019) and around 914,600 registered Syrian refugees at the end of 2019 (UNHCR 2020). Living conditions are exacerbated by a national economic collapse that led to the high devalue of the Lebanese Pound. In addition, the global COVID-19 pandemic posed further stresses on the limited resources of the national health system. All of these struggles in Lebanon were lastly aggravated by a deadly blast on the fourth of August 2020 which overwhelmed and damaged medical facilities exposing the Lebanese community in Beirut to many vulnerabilities and insecurities (UN 2020).

Lebanese adolescents are particularly vulnerable to develop mental health and substance use disorders. A study conducted on 510 adolescents aged 11–17 years in Beirut shows a 26.1% prevalence of psychiatric disorders, mainly anxiety disorders and attention-deficit/hyperactivity disorder (ADHD); however, only 6% of those with disorders reported seeking professional assistance (Maalouf et al. 2016). Followed by a secondary analysis, the study found that 4.3% of this adolescent group had suicidal experiences (Baroud et al. 2019).

According to the Ministry of Public Health (MoPH), only 5% of total financial expenditure on health is directed toward mental health (2015). There are 5 operational mental health hospitals in Lebanon with the availability of 1.26 psychiatrists and 3.42 psychologists per 100,000 population. On the other hand, half of the population remains financially undercovered for mental health services (MoPH 2015). Within this context, MoPH developed a 5-year strategy targeting prevention, promotion, and treatment of mental health and substance use disorders. Adolescents are addressed as one of the main vulnerable groups in this strategy. The strategy covers five domains: leadership and governance for mental health; reorientation and scaling up of mental health services; promotion and prevention activities for mental health and substance use disorders; information, evidence, and research; and vulnerable groups.

One of the strategic objectives is to enhance the evidence-based approach to improve mental health psychosocial support services for adolescents given the limited research in this area. Moreover, the strategy aims to scale up mental health and social services and guarantee equal access to these services. Regarding governing mental health, the strategy focuses on creating a sustainable department for mental health and substance use to allow for multi-sectoral health policy development and planning.

The Kingdom of Saudi Arabia

About half of the population of the KSA is under the age of 30, which means that, like the aforementioned Arab nations, it is a country driven by its youth (*Religion* 2020). Compared to many other Arab nations, the KSA differs in its access and quality to a higher standard of living. For example, the health expenditure per capita for the KSA in 2014 was about $18,000; in that same year, the USA's health expenditure per capita was about $9000 (*Health Expenditure* Per Capita*—1991-2014;* Health spending 2020). With more than roughly double the number of monetary resources being dedicated to healthcare spending per person in comparison to the USA, the KSA evidently has the resources necessary to provide adequate primary and preventative care to its citizens.

It is no doubt that cost and quality of care serve as major barriers to overall well-being and basic health knowledge in the KSA. According to a comprehensive review of adolescent health in Gulf countries, adolescents in the KSA mainly face noncommunicable diseases and health conditions due to lifestyle choices (Al Makadma 2017). Examples of such situations are unhealthy eating and drinking behaviors through opting for fatty, salty, and sugary items; motor vehicle accidents; and early onset of chronic and widespread tobacco smoking.

2 Part 2: What Works, What Doesn't in the Health Policies of Adolescents in the Middle East and North Africa, How the COVID-19 Impacted on Adolescence, and How Do We Strategize from Needs and Priorities lens

2.1 General Discussion: A Perspective on the Past and Current Scenes

In the past, there was a global gap in adolescent health policies. Adolescents were not targeted as a significant age group with specific health needs and priorities. Many historical declarations emphasized the rights of the child and human rights, and thus adolescents were subsequently considered under such decelerations. The term adolescence became popular in the early 1900s in Western societies (Furstenberg 2001). In the past century, the field of adolescent medicine developed, and health professionals became particularly interested in the healthcare of adolescents. Such a shift toward an adolescent-focused medicine was a result of a series of scientific advancements as well as societal changes (Alderman et al. 2003). Gradually, the world started to realize the importance of the stage of adolescence as part of societies, health systems, as well as development.

The United Nations organizations including the WHO and UNICEF addressed the issues of adolescents in the international community, and they are currently leading global agendas and strategies that target adolescents worldwide. The MDGs during 2000 and 2015 contextualized the importance of adolescent age in development agendas, and this showed that developing countries are lagging based on adolescent health-related indicators. Following the MDGs, the SDGs also address some dimensions of adolescent health and adolescent education as well as other social and economic dimensions that impact adolescents. The MENA region has been progressing toward the SDGs, and many countries integrated adolescent targets in their national health policies and strategies. In addition, some countries have specific adolescent health policies that cater to the needs of the national adolescent population.

Generally, there are gaps in addressing mental health issues in many countries of the MENA region. This could be due to cultural and religious constraints as well as a lack of professional capacities and governmental facilities. Also, there is a lack of coordination among Arab countries to develop regional goals and strategies to improve adolescent health in the region. Such initiatives are beneficial to strengthen capacities and share experiences to improve the health of the regional adolescent population. In the national context, there are different challenges on the country level given the heterogeneity of the MENA.

In the MENA region, there are some efforts and initiatives that tackle improving adolescent health. National health policies, strategies, and plans express a country's vision and define its priorities, budgetary decisions, and course of action for improving and maintaining the health of its people. The national level is a key component to deliver, monitor, and evaluate interventions that include health service reorganization, employment and training, law and policing changes, social marketing, and socioeconomic change.

As presented previously in the case studies, it is a small capture on how on a national level countries started to work on that issue. In each case study, different policies and strategies are tackling different thematic areas of adolescent mental health based on the context and needs. In this chapter four case studies were presented representing four different national policies and strategies: Sudan, Palestine, Lebanon, and the KSA.

Sudan as the first country in the region that adopted the AA-HA guidance in the adolodence policy and implementation. Through this initiative, it aims to meet the needs of adolescents and based on that its developed national strategy (WHO 2019). The strategy focuses on four components: (1) adolescent SRH, (2) violence against adolescents, (3) adolescent nutrition, and (4) unintentional injuries and the mental health of adolescents (WHO 2019). Palestine's case study was a great learning example of how stakeholders cooperated on working on an initiative to provide youth services that focus on adolescent health. This can highlight the strength of youth policy in Palestine as the establishment of the Youth Peer Education Network involved more than 500 NGOs. Talking about Lebanon, the government has already addressed adolescents in its 5-year strategy targeting prevention, promotion, and treatment of mental health and substance use disorders, where adolescents are categorized as one of the main vulnerable groups. The strategy covers five domains: Domain 1 (leadership and governance), Domain 2 (reorientation and scaling up of mental health services), Domain 3 (promotion and prevention), Domain 4 (information, evidence, and research), and Domain 5 (vulnerable groups) (MoPH 2015). For the KSA, despite realities of prosperity and relatively high quality of adolescence healthcare, unhealthily lifestyle and behavior, low health knowledge, and lack of social participation remain key issues that need to be addressed through designing appropriate national programs considering advancing the legislations, building information system, and raising awareness and healthcare practices.

Despite the great initiative in starting with addressing adolescent needs in the Arab countries, we can still find gaps in the implementation of each policy represented by the four countries. There are no statistics about the impact of the Sudanese strategy; we can say that the Sudanese adolescent strategy lacks the implementation of monitoring and evaluation framework. Unfortunately, the lack of resources could be one of the reasons for not being able to measure the impacts. Based on the situation report in Sudan for 2019, new needs for adolescents are identified especially after the malaria epidemic (OCHA 2020). This could shape the implementation of the strategy since the component of adolescent health for communicable disease prevention is not part of the four components.

The complexity of Lebanon with the political instability and the economic crises is a struggle by itself. Unfortunately, after the national economic collapse and security instability of the country, opportunities for adolescence became minimal. In addition to that, the COVID-19 pandemic posed further stresses in harsh times for Lebanon, after the revolution in late 2019, on the limited resources of the national health system. Priorities of the countries are reshaped especially after the deadly blast on August 4, 2020, which resulted in extreme loss and damages to the Beirut city and the country, especially to vulnerable people (UN 2020).

On the other hand, Palestine has similar conditions as Lebanon. Living under the Israeli occupation, adolescent people in Palestine face many struggles in living their lives due to the stressors, fragility, and violence that they live in (UNFPA 2020b). These circumstances could impact the benefits of the youth centers and services provided to adolescence. Insecurity and life instabilities are the main challenges and issues behind deteriorating adolescence life and the future in Palestine.

We can find that adolescent needs in the four countries are roughly similar with disparities, which emphasizes prioritization. Each case study focused on specific factors, which reflects the disparity of the context for each country and for the adolescent sub-population that resides in each country. Naturally, some challenges can hinder or change strategy focus. One key component is the availability of money and expertise. The current situation of child and adolescent mental health in the region is particularly challenging, given that many barriers influence any future program or policy. Adolescence challenges are preventable and must be well-planned in each country's agenda. A monitoring and evaluation plan should be addressed and implemented to evaluate if targets that were earlier set are reached.

Countries are working on frameworks, policies, projects, and initiatives that address the needs of adolescents. This is a key step for improving the future of adolescents where needs are met and services are sustained. Countries must continue working on these priorities and never retreat, as adolescents will become youth and the future of the full country will be in their hands. Adolescents need economic opportunity, safe communities, and a chance to have a voice in their future.

Adolescents Amid the Coronavirus Pandemic

Globally, adolescents face unique challenges amid the COVID-19 pandemic, which may lead to concerning circumstances such as chronic stress and unhealthy coping mechanisms, adverse mental health outcomes, long-term physical inactivity, lack of communication, and social contact, and barriers to accessing high-quality education. This global pandemic crisis has also affected most of the adolescent national, regional, and global policies, programs, and routine services. There are also various obstacles that individuals, regardless of age, ethnicity, or nationality, may face amid the COVID-19 pandemic, a public health emergency that has temporarily prohibited travel and close physical contact. The stay-at-home orders may lead to an increase in domestic abuse, food insecurity, unemployment, inadequate access to education, inadequate access to healthcare including menstrual hygiene products, and a lack of access to technological devices and/or the Internet, among many more.

Besides the psychosocial health implications of COVID-19 on adolescents, there is a growing body of literature that indicates that infection with the virus may also lead to detrimental health consequences and outcomes. A recent study involving 82 participating healthcare institutions across 25 European countries found that "individuals with viral co-infection (i.e., infected with severe acute respiratory syndrome-associated Coronavirus (SARS-CoV-2) and one or more other viral agents) were more likely to require Intensive Care Units (ICU) support than those in whom SARS-CoV-2 was the only viral agent identified. This may be exacerbated for the winter period 2020–21, when the incidence of other viral respiratory tract infections, including (Respiratory Syncytial Virus) RSV and influenza virus infections, is bound to increase" (Götzinger et al. 2020). This study, among many others similar in research design and participant population, highlights the need for greater and more impactful policies and interventions that are specific to adolescents amid and after the COVID-19 pandemic.

Psychosocial Impacts of COVID-19 on Adolescents Globally

Physical and social distancing and quarantine guidelines, while essential to minimizing COVID-19 incidence and mortality rates, have led adolescents to experience psychological distress and feel nervous, helpless, and fearful (Saurabh and Ranjan 2020). A recent study involving 1143 parents of Italian and Spanish children aged 3–18 years found that about 86% of parents detected shifts in their children's emotional states and behaviors amid quarantine. Parents specifically perceived changes in concentration, boredom, irritability, restlessness, nervousness, loneliness, and uneasiness (Orgiles et al. 2020).

Research studies focusing on the psychosocial impacts of COVID-19 on adolescents globally have also been detecting symptoms of anxiety and depression among certain populations. A study based in China found that, of the 8079 adolescents involved in the study, 43.7%, 37.4%, and 31.3% experienced depressive symptoms, anxiety symptoms, and a combination of depressive and anxiety symptoms, respectively (Zhou et al. 2020). This study also found that depressive and anxiety symptoms were more likely to be experienced by adolescents with poor knowledge of COVID-19. This finding supports that increased educational and social interventions such as philanthropic and voluntary initiatives, recreational programs, and health education summer programs may be necessary to aid with mental health challenges that adolescents are facing.

Adolescence, across cultural and religious groups, is a time of great development in terms of self-independence and awareness. In addition to the impacts of COVID-19 on physical and mental health, adolescents have faced severe changes in the quality of their education. According to the World Bank, as of March 24, 2020, 160 countries ordered school closures that affected over 1.5 billion children and youth. Learning and technology poverty in the form of school closures may cause not only loss of learning in the short term but also further loss in human capital and diminished economic opportunities over the long term. A global report qouted "as seen from previous health emergencies, most recently the Ebola outbreaks, the impact on education is likely to be most devastating in countries with already low learning outcomes, high dropout rates, and low resilience to shocks," (The World Bank, 2020). This finding, among all studies and topics previously addressed, reveals that adolescents are facing adverse physical and mental health, educational, and developmental challenges amid COVID-19. Producing and implementing interventions that can help minimize these issues is essential because the duration of the pandemic is still unknown and current challenges may cause long-term effects that remain even after a vaccine and cure are found for COVID-19.

The Needs and Priorities of Adolescents at the Policy Level in the MENA Region

According to UNICEF, "Children and young people (0 to 24 years) in the MENA currently account for nearly half of the region's population and have the potential to become agents of change, contributing to a more prosperous and stable future for themselves and their communities, and playing their part in reaping the demographic dividend." The WHO has also identified youth unemployment, gender gaps, war, and sociopolitical conflict, and education and health inaccessibility to be especially pressing issues for adolescents in the MENA region. While social, economic, and political conditions vary across the region, it is important to consider the impacts of gender, religion, familial situation, socioeconomic status, and forced migration as important factors that can overwhelmingly influence COVID-19-related outcomes. For example, in terms of education, only 62% of Arab youth aged 15–29 years have Internet access in their community and just 22% have access at home (Eastern Mediterranean Health Journal 2015). The OECD reports that "In the MENA region, more than 110 million school-aged students have limited access to education (formal and non-formal) due to closure of school and university facilities, youth centers and other public spaces" (OECD 2020). Additionally, many low-income individuals and/or survivors of forced migration may not be able to afford access to education or virtual resources.

Employment opportunities for this population were already limited pre-pandemic, which means that those who were essential workers may be left jobless or working in unsafe conditions. A recent report explained that "The economic outlook for these disadvantaged populations is poor, as estimated by the UNESCWA: approximately 1·seven million people in the region will most probably lose their jobs in 2020, and eight million more people will fall into poverty, half of whom will be children" (Karamouzian and Madani 2020). The MENA region holds the highest youth unemployment rate in the world and this situation is bound to worsen amid widespread jobless due to COVID-19.

2.2 *Conclusion and Future Roadmap for Policy Development of Adolescent Health in MENA*

This chapter addresses central public health questions that require further attention by MENA health policymakers. Through extracting data from primary and secondary sources, thematically exploring them, and synthesizing an in-depth analysis to report on the main findings, this chapter adds to a growing body of literature on global and regional health policy amid the COVID-19 pandemic. Adolescents are currently facing unprecedented changes that have resulted in social, health, political, and economic impacts globally due to the COVID-19 pandemic. Various social determinants of health may influence COVID-19 incidence and mortality rates, as well as adverse psychosocial and behavioral outcomes. In the MENA region, the needs and priorities of adolescents include responses to severe unemployment, gender gaps, war and sociopolitical conflict, and education and health inaccessibility. These issues will only worsen amid the COVID-19 pandemic, a tragic reality that should urge federal and local governments to adequately respond to such conditions. Many countries in the region are on their track to achieve the SDGs related to adolescent health; however, this progress gets interrupted due to instability, social unrest, or epidemics such as the COVID-19 outbreak. Two categories of countries were observed in terms of health policies for adolescence and their services tend to be as follows: high-resource countries are stepping forward by formulating, adopting, and implementing effective policies and programs that provide optimal adolescent services, while other less-resourced countries are facing constraints such as capacity, resources, and commitment. Considerable efforts were and are being exerted in the region to address adolescent health in national health policies within the context of comprehensive sustainable development. Stakeholder involvement requires more attention in formulating policies and designing national adolescent health programs. New approaches such as an adolescent-focused healthcare systems, evidence-based adolescent advocacy and policies, and adolescent health-related indicators were not consistently applied in these global contexts. This chapter provides a summary of COVID-19-related outcomes on adolescents globally and in the MENA region, as well as recommendations to help improve current conditions in the MENA region specifically.

2.3 *Public Health Policy and Intervention Recommendations in the MENA Region*

While many nations have found difficulty in proactively implementing COVID-19-specific policies and interventions for adolescents, various countries in the MENA region that were already facing poverty, conflict, and humanitarian crises are especially struggling. It is important to address, brainstorm, and suggest recommendations for policies and interventions in the MENA region. Such ideas may include:

1. Increased education and awareness of adolescents' rights and services as well as current challenges such as COVID-19 (what it is, how to minimize the spread, how to remain healthy psychologically and socially, and local resources) to combat misinformation that is mostly spread via social media.
2. Increased stimulus economic support and checks, especially for impoverished individuals who have been left jobless or whose jobs pose a grave risk for COVID-19 transmission.
3. More confidential and accessible telemedicine care and digital consultations (especially access to SRH).
4. Virtual therapy and mental health support (including free text services for those who cannot talk due to lack of privacy at home).
5. Better quality remote education (provide students with necessary technological devices and Internet access, and train teachers to teach in more engaging and effective virtual methods).
6. Partnerships between NGOs and local governments to work with refugee and internally displaced people (IDP) camps to implement the aforementioned suggestions in culturally and religiously appropriate ways.
7. Communication between federal governments, public health experts and associated stakeholders, and the international community to guarantee essential routine services sustained and needs fulfilled, including those related to COVID-19 vaccines and cures, once they become available, in the MENA region (including access for low-income individuals and survivors of forced migration).
8. Adolescence health programs and services should be integrated in the primary healthcare system, and all stakeholders, including adolescence organizations, groups, and initiatives, must be engaged in efforts of needs assessment, strategizing, prioritization, and policies designing.
9. New and innovative approaches, concepts, and principles such as UHC, Adolescent-Focused Healthcare System (AFHS), Evidence-Based Adolescence Advocacy and Policies (EBAP), System-Wide Approach (SWA), Human

Rights-Based Approach (HBRA), and Adolescent Health-Related Indicators (AHRI) should be adopted and applied at the institutional and national levels.
10. The existing legislative frameworks and regulations need to be revisited by all actors in many countries to abreast of ever-changing and increasing needs and priorities of adolescence in the region.

2.4 *Future Outlook and Critical Questions for Consideration*

As it has been observed in past and current literature and concluded to in this comprehensive synthesis, the MENA is a region that is greatly diverse in race, ethnicity, religion, class, sociopolitical status, history, national values, and projected trajectory. Each nation, depending on the factors formerly listed, may or may not have the capacity and resources to evaluate existing public and clinical health outcomes, strategically plan for potential solutions, and collaborate with interdisciplinary teams and stakeholders to address major gaps. This reality has resulted in variance regarding the status of adolescent health, where some nations simply are not in good social, political, or economic standing to even address adolescent health, while others have completely re-defined the scope and services of adolescent health to be more socially and religiously competent to local values. The nature of adolescent health, both as a privilege and a malleable field of medicine, prompts critical questions that must be considered by state and non-state sectors in three fundamental avenues, practice, research, and education:

1. How can adolescent health be defined and/or standardized across the MENA? Would a regional adolescent health strategy or framework help advance adolescent health policy in the region's countries?
2. Should and how can countries who are currently facing sociopolitical turmoil be expected to work on an agenda addressing adolescent health needs and priorities?
3. What role can various digital health technologies play in the development and progression of adolescent health across the MENA?
4. Will the COVID-19 pandemic serve as a detriment to the advancement of adolescent health in the region and will the pandemic impose all stakeholders in the region to change the mindset that run adolescent health programs planning and implementation? And how can existing infrastructure and temporary COVID-19 resources and guidelines be pivoted to support adolescents and their families?
5. What is the role of scientific research in the adolescent health in the MENA region and does its services and programs driven by the approach of evidence-based practice?
6. What short- and long-term goals are important to highlight across the MENA region in terms of health advancement on the system level? What types of health policy programs and services are of utmost importance to sustain and support in the MENA region?

7. How can other health sectors and systems such as public and clinical health support the field of health policy? And what stakeholders, outside the health sector, are important to be incorporated in health policy formulation and implementation?

References

Abdul Rahim HF, Sibai A, Khader Y, Hwalla N, Fadhil I, Alsiyabi A, et al. (2014). Noncommunicable diseases in the Arab world. Lancet; 383:356-67.

Al Makadma AS (2017) Adolescent health and health care in the Arab Gulf countries: Today's needs and tomorrow's challenges. *International Journal of Pediatrics and Adolescent Medicine* 4(1), 1–8

AlBuhairan, Fadia (2015). Adolescent and Young Adult Health in the Arab Region: Where We Are and What We Must Do. *Journal of Adolescent Health.* Vol.57, Issue 3, pp.249-251. https://doi.org/10.1016/j.jadohealth.2015.06.010

Alderman, E., Rieder, J. & Cohen, M. (2003).The History of Adolescent Medicine. Pediatric Research 54, 137–147). https://doi.org/10.1203/01.PDR.0000069697.17980.7C

AlKhaldi, M., Alkaiyat, A., Abed, Y. et al. The Palestinian health research system: who orchestrates the system, how and based on what? A qualitative assessment. Health Res Policy Sys 16, 69 (2018). doi:https://doi.org/10.1186/s12961-018-0347-4

AlKhaldi, M., Meghari, H., and Ahmed, S. A New Era for Health Policy and Systems Research. *Think Global Health.* July 20, 2020a. https://www.thinkglobalhealth.org/article/new-era-health-policy-and-systems-research

AlKhaldi, M., Kaloti, R., Shella, D., Al Basuoni, A., & Meghari, H. (2020b). Health system's response to the COVID-19 pandemic in conflict settings: Policy reflections from Palestine. *Global Public Health, 15*(8), 1244-1256. https://doi.org/10.1080/17441692.2020.1781914

Baroud, E., Ghandour, L.A., Alrojolah, L., Zeinoun, P., and Maalouf, F.T. (2019). Suicidality among Lebanese adolescents: Prevalence, predictors and service. Utilization. *Psychiatry Research* 275 338–344. https://doi.org/10.1016/j.psychres.2019.03.033

Belaid, L., Bayo, P., Kamau, L. Nakimuli, E, Omoro, E. Lobor, R. Samson, B. Dimiti, A (2020). Health policy mapping and system gaps impeding the implementation of reproductive, maternal, neonatal, child, and adolescent health programs in South Sudan: a scoping review. Confl Health 14, 20 (2020). doi:https://doi.org/10.1186/s13031-020-00258-0

Daher-Nashif, S., and Bawadi, H. (2020). Women's Health and Well-Being in the United Nations Sustainable Development Goals: A Narrative Review of Achievements and Gaps in the Gulf States. *International Journal of Environmental Research and Public Health* 17, 3: 1059. https://doi.org/10.3390/ijerph17031059

DeJong J., and Bashour H. (2016). Regional Report: Sexual and Reproductive Health Laws and Policies in Selected Arab Countries. UNFPA, Middle East and North Africa Health Policy Forum.

Fehling M., Jarrah Z.M., Tiernan M.E., Albezreh S., VanRooyen M.J., Alhokair A., and Nelson B.D. (2015). Youth in crisis in the Middle East and North Africa: a systematic literature review and focused landscape analysis. Eastern Mediterranean Health Journal. Vol. 21 No. 12

Ferguson J, Dick B. (2012). Policies to support adolescents' health. Health for the World's adolescents: a second chance in the second decade. Geneva, Switzerland: World Health Organization. Accessed 4 July 2020. Retrieved from: http://apps.who.int/adolescent/second-decade/section8.

F.F. Furstenberg. (2001). Adolescence, Sociology of, Editor(s): Neil J. Smelser, Paul B. Baltes, International Encyclopedia of the Social & Behavioral Sciences, Pergamon, 2001,Pages 94-97, ISBN 9780080430768, https://doi.org/10.1016/B0-08-043076-7/01994-X.

Götzinger, F., Santiago-García, B., Noguera-Julián, A., Lanaspa, M., Lancella, L., Calò Carducci, F. I., Gabrovska, N., Velizarova, S., Prunk, P., Osterman, V., Krivec, U., Lo Vecchio, A., Shingadia, D., Soriano-Arandes, A., Melendo, S., Lanari, M., Pierantoni, L., Wagner, N., L'Huillier, A. G., Riordan, A. (2020). COVID-19 in children and adolescents in Europe: A multinational, multicentre cohort study. *The Lancet Child & Adolescent Health*, S2352464220301772. https://doi.org/10.1016/S2352-4642(20)30177-2

Hamdan M. and Imam A. (2019). Mapping Adolescent and Youth Sexual and Reproductive Health Services in Palestine. Palestinian Medical Relief Society (PMRS).

Hamad Sh., Abu Hamra E., Diab R., Abu Hamad B., Jones N., and Małachowskal A. (2020). Exploring the impacts of COVID-19 on adolescents in the Gaza Strip. Listening to young people's voices under COVID-19.

Health spending. (2020.). [Data set]. OECD. https://doi.org/10.1787/8643de7e-en

Karamouzian, M., & Madani, N. (2020). COVID-19 response in the Middle East and north Africa: Challenges and paths forward. *The Lancet Global Health*, *8*(7), e886–e887. https://doi.org/10.1016/S2214-109X(20)30233-3

Kuruvilla, S., Bustreo, F. et al. (2016). The Global strategy for women's, children's, and adolescents' health (2016–2030): a roadmap based on evidence and country experience. Bull World Health Organ. 94(5): 398–400. https://doi.org/10.2471/BLT.16.170431

Maalouf, F.T., Ghandour, L.A., Halabi, F., Zeinoun, P., Shehab, A.A.S. and Tavitian, L. (2016). Psychiatric disorders among adolescents from Lebanon: prevalence, correlates, and treatment gap. *Social Psychiatry and Psychiatric Epidemiology*, 51(8), 1105–1116. https://doi.org/10.1007/s00127-016-1241-4

Ministry of Public Health (MoPH) (2015). Mental Health and Substance Use: Prevention, Promotion, and Treatment. Situation Analysis and Strategy for Lebanon. Available at: https://www.moph.gov.lb/userfiles/files/Mental%20Health%20and%20Substance%20Use%20Strategy%20for%20Lebanon%202015-2020-V1_1-English.pdf (Accessed 10 August 2020).

Ministry of Health (2019). Health Annual Report, Palestine, 2018. Palestinian Health Information Center (PHIC) -MOH.

OCHA (2020). Humanitarian Response Plan Sudan. Humanitarian Programme Cycle 2020. https://www.who.int/health-cluster/countries/sudan/Sudan-Humanitarian-Response-Plan-January-2020.pdf?ua=1

OECD Development Centre (2018). "Youth Well-being Policy Review of Jordan", EU-OECD Youth Inclusion Project, Paris.

OECD Policy Responses to Coronavirus (COVID-19) COVID-19 crisis response in MENA countries Updated 06 November 2020. https://www.oecd.org/coronavirus/policy-responses/covid-19-crisis-response-in-mena-countries-4b366396/?fbclid=IwAR2RR_NSBDdVLMlz4pzEWBOGsPuPL3HmZ1doLHrfHCgcP5mAwi-EX8qnjbI

Office for the Coordination of Humanitarian Affairs (OCHA), 2019. Sudan situation report 12 Dec 2019. Accessed by 7th July 2020: https://reliefweb.int/report/sudan/sudan-situation-report-12-dec-2019-enar

Orgiles, M., Morales, A., Delveccio, E., Mazzeschi, C., & Espada, J. P. (2020). Immediate Psychological Effects of COVID-19 Quarantine in Youth from Italy and Spain. *SSRN Electronic Journal*. https://doi.org/10.2139/ssrn.3588552

Palestinian Central Bureau of Statistics (2018). General Population, Housing and Establishments Census 2017: Summary of the final results of the census. Ramallah - Palestine.

Patton, G. C., Sawyer, S. M., Santelli, J. S., Ross, D. A., Afifi, R., Allen, N. B., Arora, M., Azzopardi, P., Baldwin, W., Bonell, C., Kakuma, R., Kennedy, E., Mahon, J., McGovern, T., Mokdad, A. H., Patel, V., Petroni, S., Reavley, N., Taiwo, K., Waldfogel, J., … Viner, R. M. (2016). Our future: a Lancet commission on adolescent health and wellbeing. *Lancet (London, England)*, *387*(10036), 2423–2478. doi:https://doi.org/10.1016/S0140-6736(16)00579-1

Saurabh, K., & Ranjan, S. (2020). Compliance and Psychological Impact of Quarantine in Children and Adolescents due to Covid-19 Pandemic. *The Indian Journal of Pediatrics*, *87*(7), 532–536. https://doi.org/10.1007/s12098-020-03347-3

Sayej S., and Qtait M. (2016). Adolescence (13-19yrs) in Palestinian. Vol.12. Journal of Pharmacy and Alternative Medicine.

UN Committee on the Rights of the Child (CRC), General Comment No. 20 (2016) *on the implementation of the rights of the child during adolescence*, 6 December 2016, CRC/C/GC/20, available at https://www.refworld.org/docid/589dad3d4.html [accessed 10 July 2020]

UNFPA (2018) regional intervention action plan for Arab states 2018-2021; accessed by 26th May 2020 https://www.unfpa.org/sites/default/files/admin-resource/Final_RIAP_ASRO.pdf

United Nations (n.d.). Global Issue, Youth. Available at https://www.un.org/en/sections/issues-depth/youth-0/

United Nations (UN) (2015). *The Millennium Development Goals Report 2015*. Available at: https://www.un.org/millenniumgoals/2015_MDG_Report/pdf/MDG%202015%20rev%20(July%201).pdf (Accessed 26 May 2020).

United Nations, Department of Economic and Social Affairs, Population Division (2019). World Population Prospects 2019: Highlights (ST/ESA/SER.A/423). https://reliefweb.int/sites/reliefweb.int/files/resources/WPP2019_Highlights.pdf

United Nations (UN) (2020). Beirut blast: Here's how you can help the UN aid Lebanon's recovery. UN News 7 August 2020. Available at: https://news.un.org/en/story/2020/08/1069782 (Accessed 9 August 2020).

United Nations International Children's Emergency Fund (UNICEF 2018). MENA generation 2030: investing in children and youth today to secure a prosperous region tomorrow: accessed 18th May 2020: https://www.unicef.org/mena/reports/mena-generation-2030

United Nations Inter-Agency Technical Task Team on Young People (UN IATTTYP) (2018). Regional Framework of Joint Strategic Actions for Young People 2016-2017 in the Arab States and the Middle East and North Africa Region. Accessed 20th May 2020 https://documents.wfp.org/stellent/groups/public/documents/op_reports/wfp282333.pdf

United Nations Children's Fund (UNICEF) (2019). *MENA Generation 2030: Investing in children and youth today to secure a prosperous region*. Available at: tomorrow https://www.unicef.org/mena/media/4141/file/MENA-Gen2030.pdf (Accessed 2 June 2020).

United Nations Economic and Social Commission for Western Asia (UNESCWA) (2020). *Arab Sustainable Development Report 2020*. Chapter 3: SDG 3 Ensure healthy lives and promote well-being for all at all ages. Available at: https://asdr.unescwa.org/sdgs/pdf/en/ASDR2020-SDGs/ASDR2020-SDG3.pdf. (Accessed 13 June 2020).

United Nations Population Fund (2014). Independent Country Programme Evaluation: Lebanon 2010–2014. Evaluation Office, 605 Third Avenue, New York, NY 10158 U.S.A.

United Nations Population Fund (2020a). Young people. Accessed 4 June 2020. Retrieved from: https://jordan.unfpa.org/en/node/22551

United Nations Population Fund (2020b). Young people. Accessed 4 June 2020. Retrieved from: https://palestine.unfpa.org/en/node/22580

United Nations Statistics Division (UNSD) (2020). *A global indicator framework for the Sustainable Development Goals and targets of the 2030 Agenda for Sustainable Development.* Global indicator framework adopted by the General Assembly (A/RES/71/313), annual refinements contained in E/CN.3/2018/2 (Annex II), E/CN.3/2019/2 (Annex II), and 2020 Comprehensive Review changes (Annex II) and annual refinements (Annex III) contained in E/CN.3/2020/2. *Available at*: https://unstats.un.org/sdgs/indicators/Global%20Indicator%20Framework%20after%202020%20review_Eng.pdf (Accessed 27 May 2020).

United Nations Relief and Works Agency for Palestinian Refugees in the Near East (UNRWA) (2019). *Where We Work: Lebanon*. Available at: https://www.unrwa.org/where-we-work/lebanon (Accessed 2 August 2020).

United Nations High Commissioner for Refugees (UNHCR) (2020). *2019 Year-End report, Operation Lebanon*. Available at: https://reporting.unhcr.org/sites/default/files/pdfsummaries/GR2019-Lebanon-eng.pdf (Accessed 2 August 2020).

WHO (2014). Health for the World's adolescents: a second chance in the second decade. Geneva, Switzerland: World Health Organization. Accessed 1st June 2020. Retrieved from: https://apps.who.int/adolescent/second-decade/files/1612_MNCAH_HWA_Executive_Summary.pdf

WHO (2016). The Global AA-HA Framework (Accelerated Action for the Health of Adolescents). Towards implementation of the Global Strategy for Women's, Children's, and Adolescents' Health (2016-2030).

WHO (2017) Global Accelerated Action for the Health of Adolescents (AA-HA!): guidance to support country implementation. Geneva: World Health Organization; License: CC BY-NC-SA 3.0 IGO.

WHO (2019) Lessons learned from applying the Accelerated Action for the Health of Adolescents (AA-HA!) guidance for policy development in early adopter countries. Sudan. Geneva: World Health Organization; 2019. License: CC BY-NC-SA 3.0 IGO

WHO (2020). WHO resolutions. Accessed 8th May 2020: https://www.who.int/medical_devices/policies/who_resolutions/en/

WHO EMRO | *Youth in crisis in the Middle East and North Africa: A systematic literature review and focused landscape analysis | Volume 21, issue 12 | EMHJ volume 21, 2015*. http://www.emro.who.int/emhj-volume-21-2015/volume-21-issue-12/youth-in-crisis-in-the-middle-east-and-north-africa-a-systematic-literature-review-and-focused-landscape-analysis.html. (Accessed 8 August 2020)

WHO Regional Office for the Eastern Mediterranean (2019) WHO in the Eastern Mediterranean Region: annual report of the Regional Director 2018. Cairo. Licence: CC BY-NC-SA 3.0 IGO

World Bank Education and COVID-19. (2020). Retrieved, from https://www.worldbank.org/en/data/interactive/2020/03/24/world-bank-education-and-covid-19. (Accessed 8 August 2020)

World Health Organization (2018) Adolescent health: the missing population in universal health coverage. Accessed 23rd May 2020: https://www.who.int/pmnch/media/news/2018/Adolescent-Health-Missing-Population-in-UHC.pdf?ua=1

World Health Organization (n.d.). Adolescent health: Adolescent health in the South-East Asia Region. Available at: https://www.who.int/southeastasia/health-topics/adolescent-health

World Health Organization (WHO) (2015). *Health in 2015 from MDGs to SDGs*. Available at: https://apps.who.int/iris/bitstream/handle/10665/200009/9789241565110_eng.pdf. (Accessed 25 June 2020)

World Health Organization (WHO) (2019). Health Conditions in the Occupied Palestinian Territory, including East Jerusalem, and in the Occupied Syrian Golan: Report by the Director-General. Available at: https://apps.who.int/iris/bitstream/handle/10665/328758/A72_33-en.pdf?sequence=1&isAllowed=y. (Accessed 21 June 2020).

Zhou, S.-J., Zhang, L.-G., Wang, L.-L., Guo, Z.-C., Wang, J.-Q., Chen, J.-C., Liu, M., Chen, X., & Chen, J.-X. (2020). Prevalence and socio-demographic correlates of psychological health problems in Chinese adolescents during the outbreak of COVID-19. *European Child & Adolescent Psychiatry, 29*(6), 749–758. doi:https://doi.org/10.1007/s00787-020-01541-4

Index

C. Barakat, F. Al Anouti (eds.), *Adolescent Mental Health in The Middle East and North Africa*, Global Perspectives on Health Geography,
https://doi.org/10.1007/978-3-030-91790-6

The manufacturer's authorised representative in the EU is Springer Nature Customer Service Centre GmbH, Europaplatz 3, 69115 Heidelberg, Germany. If you have any concerns regarding our products, please contact ProductSafety@springernature.com

Printed and bound by CPI Group (UK) Ltd, Croydon, CR0 4YY

30/09/2026

02225265-0005